SPECIAL ISSUE OF // THE 2013 DIALOGUE TO THE WORLD PENJING //

我们也

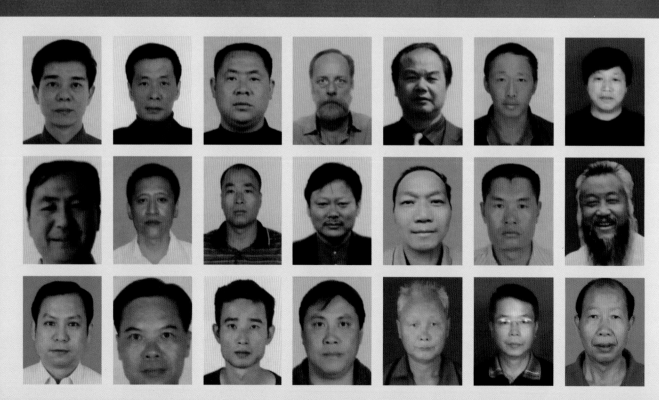

来了 WE
ARE COMING TOO.

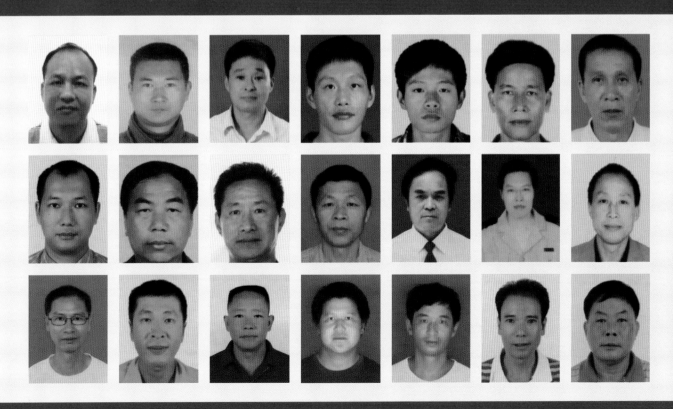

中国盆景赏石

2013 "唐苑的世界盆景对话" 特别专辑
Special Issue of "the 2013 Dialogue to the World Penjing"

中国林业出版社 China Forestry Publishing House

向世界一流水准努力的
——中文高端盆景媒体

《中国盆景赏石》

世界上第一本全球发行的中文大型盆景媒体
向全球推广中国盆景文化的传媒大使
为中文盆景出版业带来全新行业标准

《中国盆景赏石》
2012年1月起
正式开始全球（月度）发行

图书在版编目（CIP）数据

中国盆景赏石. 2013. "唐苑的世界盆景对话" 特别专辑 / 中国盆景艺术家协会主编. -- 北京：中国林业出版社，2014.2
ISBN 978-7-5038-7380-5

Ⅰ.①中… Ⅱ.①中… Ⅲ.①盆景－观赏园艺－中国－丛刊②石－鉴赏－中国－丛刊 Ⅳ.①S688.1-55 ②TS933-55

中国版本图书馆 CIP 数据核字 (2014) 第 023689 号

责任编辑：何增明 张华
出 版：中国林业出版社
　　　　E-mail:shula5@163.com
　　电话：(010) 83286967
社 址：北京西城区德内大街刘海胡同7号
　　邮编：100009
发 行：中国林业出版社
印 刷：北京利丰雅高长城印刷有限公司
开 本：230mm×300mm
版 次：2014年2月第1版
印 次：2014年2月第1次
印 张：8
字 数：200千字
定 价：48.00元

主办、出品、编辑： 中国盆景艺术家协会

E-mail：penjingchina@yahoo.com.cn
Sponsor/Produce/Edit: China Penjing Artists Association

创办人、总出版人、总编辑、视觉总监、摄影：苏放
Founder, Publisher, Editor-in-Chief, Visual Director, Photographer: Su Fang
电子邮件：E-mail：1440372565@qq.com

《中国盆景赏石》荣誉行列——集体出版人（以姓氏笔画为序、按月轮换）：
胡世勋、柯成昆、谢克英、曾安昌、樊顺利、黎德坚、魏积泉、于建涛、王礼宾、申洪良、刘常松、刘传刚、刘永洪、汤锦铭、李城、李伟、李正银、芮新华、吴清昭、吴明选、吴成发、陈明兴、罗贵明、杨贵生

《中国盆景赏石》高级顾问团队（以姓氏笔画为序、按月轮换）：
曾尔恩、李晓、吴国庆、陈伟、夏敬明、黄金耀、曹志振

《中国盆景赏石》顾问团队（以姓氏笔画为序、按月轮换）：
储朝辉、关山、李晓波、陈家荣

名誉总编辑Honorary Editor-in-Chief: 苏本一 Su Benyi
名誉总编委Honorary Editor: 梁悦美 Amy Liang
名誉总顾问Honorary Advisor: 张世藩 Zhang Shifan

美术总监Art Director: 杨竞Yang Jing
美编Graphic Designers: 杨竞Yang Jing 杨静Yang Jing 尚聪Shang Cong 李锐Li Rui
摄影Photographer: 苏放Su Fang 纪武军Ji Wujun
总编助理Assistant of Chief Editor: 徐雯Xu Wen
编辑Editors: 雷敬敷Lei Jingfu 孟媛Meng Yuan 霍佩佩Huo Peipei 苏春子Su Chunzi 房岩Fang Yan

编辑报道热线：010-58693878（每周一至五：上午10:00-下午6:30）
News Report Hotline: 010-58693878 (10:00a.m to 6:30p.m, Monday to Friday)
传真Fax：010-58693878
投稿邮箱Contribution E-mail：CPSR@foxmail.com
会员订阅及协会事务咨询热线：010-58690358（每周一至五：上午10:00-下午6:30）
Subscribe and Consulting Hotline: 010-58690358 (10:00a.m to 6:30p.m, Monday to Friday)
通信地址：北京市朝阳区建外SOHO16号楼1615室 邮编：100022
Address: JianWai SOHO Building 16 Room 1615, Beijing ChaoYang District, 100022 China

编委Editors（以姓氏笔画为序、按月轮换）：曹志振、储朝辉、于建涛、王礼宾、王选民、申洪良、刘常松、刘传刚、刘永洪、汤锦铭、关山、李城、李伟、李正银、李晓、李晓波、张树清、芮新华、吴清昭、吴明选、吴成发、吴国庆、陈明兴、陈瑞祥、陈伟、陈家荣、罗贵明、杨贵生、胡乐国、胡世勋、郑永泰、柯成昆、赵庆泉、徐文强、袁心义、张乐江、谢克英、曾安昌、鲍世骐、潘仲连、樊顺利、黎德坚、魏积泉、蔡锡元、李先进、夏敬明、黄金耀、曾尔恩

中国台湾及海外名誉编委兼顾问：山田登美男、小林国雄、须藤雨伯、小泉薰、郑成燕、成范永、李仲鸿、金世元、森前诚二
China Taiwan and Overseas Honorary Editors and Advisors: Yamada Tomio, Kobayashi Kunio, Sudo Uhaku, Koizumi Kaoru, Zheng Chenggong, Sung Bumyoung, Li Zhonghong, Kim Saewon, Morimae Seiji

技术顾问：潘仲连、赵庆泉、铃木伸二、郑诚恭、胡乐国、徐昊、王选民、谢克英、李仲鸿、郑建良
Technical Advisers: Pan Zhonglian, Zhao Qingquan, Suzuki Shinji, Zheng Chenggong, Hu Leguo, Xu Hao, Wang Xuanmin, Xie Keying, Li Zhonghong, Zheng Jianliang

协办单位：中国罗汉松生产研究示范基地【广西北海】、中国盆景名城——顺德、《中国盆景赏石》广东东莞真趣园读者俱乐部、广东中山古镇绿博园、中国盆景艺术家协会中山古镇绿博园会员俱乐部、漳州百花村中国盆景艺术家协会福建会员俱乐部、南通久发绿色休闲农庄、宜兴市鉴石紫砂盆艺研究所、广东中山虫二居盆景园、漳州天福园古玩城

驻中国各地盆景新闻报道通讯站点：鲍家盆景园（浙江杭州）、"山茅草堂"盆景园（湖北武汉）、随园（江苏常州）、常州市职工盆景协会、柯家花园（福建厦门）、南京市职工盆景协会（江苏）、景铭盆景园（福建漳州）、趣怡园（广东深圳）、福建晋江鸿江盆景植物园、中国盆景大观园（广东顺德）、中华园（山东威海）、佛山市奥园置业（广东）、清怡园（江苏昆山）、樊氏园林景观有限公司（安徽合肥）、成都三邑园艺绿化工程有限责任公司（四川）、漳州百花村中国盆景艺术家协会福建会员会交流基地（福建）、同蓝园（广东东莞）、屹松园（江苏昆山）、广西北海银阳园艺有限公司、湖南裕华化工集团有限公司盆景园、海南省盆景专业委员会、海口市花卉盆景产业协会（海南）、海南鑫山源热带园林艺术有限公司、四川省自贡市贡井百花苑度假山庄、遂苑（江苏苏州）、厦门市盆景花卉协会（福建）、苏州市盆景协会（江苏）、厦门市雅石盆景协会（福建）、广东省盆景协会、广东省顺德盆景协会、广东省东莞市茶山盆景协会、重庆市星星矿业盆景园、浙江省盆景协会、山东省盆景协会、广东省大良盆景协会、广东省容桂盆景协会、北京市盆景赏石艺术研究会、江西省萍乡市盆景协会、中国盆景艺术家协会四川会员俱乐部、《中国盆景赏石》（山东文登）五针松生产研究读者俱乐部、漳州瑞祥陶艺术投资有限公司（福建）、泰州盆景研发中心（江苏）、芜湖金日矿业有限公司（安徽）、江苏丹阳兰馨盆景园艺社、晓虹园（江苏扬州）、金陵半古园（江苏南京）、龙海市华兴榕树盆景园（福建漳州）、华景园、如皋市花木大世界（江苏）、金陵盆景赏石博览园（江苏南京）、海口锦园（海南）、一口轩、天宇盆景园（四川自贡）、福建盆景示范基地、集美园林市政公司（福建厦门）、广东英盛盆景园、水晶山庄盆景园（江苏连云港）

中国盆景艺术家协会拥有本出版品图片和文字及设计创意的所有版权，未经版权所有人书面批准，一概不得以任何形式或方法转载和使用，翻版或盗版创意必究。
Copyright and trademark registered by Chinese Penjing Artists Association. All rights reserved. No part of this publication may be reproduced or used without the written permission of the publisher.

法律顾问：赵煜
Legal Counsel: Zhao Yu

制版印刷：北京利丰雅高长城印刷有限公司
读者凡发现本书有掉页、残页、装订有误等印刷质量问题，请直接邮寄到以下地址，印刷厂将负责退换：北京市通州区中关村科技园通州光机电一体化产业基地政府路2号 邮编101111
联系人王莉，电话：010-59011332。

VIEW CHINA
景色中国

黑松 *Pinus thunbergii* 高70cm 宽120cm 中国唐苑藏品 刘伟民摄影
Japan Black Pine. Height: 70cm, Width: 120cm. Collector: China Tang Yuan. Photographer: Liu Weimin

中国盆景赏石

CHINA PENJING & SCHOLAR'S ROCKS

2013 "唐苑的世界盆景对话" 特别专辑
Special Issue of "the 2013 Dialogue to the World Penjing"

封面：黑松 *Pinus thunbergii* 高 75cm 宽 90cm 中国唐苑藏品 刘伟民摄影
Cover: Japan Black Pine. Height: 75cm, Width: 90cm. Collector: China Tang Yuan. Photographer: Liu Weimin

封四："玉枕" 彩陶石 长 55cm 宽 20cm 高 9cm 李正银藏品 苏放摄影
"Jade Pillow". Painted Pottery Stone. Length: 55cm, Width: 20cm, Height: 9cm. Collector: Li Zhengyin, Photographer: Su Fang

我们也来了
We Are Coming Too

卷首语 Preamble

08 中国唐苑——一个男人的梦想 文：苏放
Preamble: China Tang Yuan-One Man's Dream Author: Su Fang

景色中国 VIEW CHINA

03 黑松 *Pinus thunbergii* 高 70cm 宽 120cm 中国唐苑藏品 刘伟民摄影
Japan Black Pine. Height: 70cm, Width: 120cm. Collector: China Tang Yuan. Photographer: Liu Weimin

12 黑松 *Pinus thunbergii* 高 95cm 宽 100cm 中国唐苑藏品 刘伟民摄影
Japan Black Pine. Height: 95cm, Width: 100cm. Collector: China Tang Yuan. Photographer: Liu Weimin

13 黑松 *Pinus thunbergii* 高 100cm 宽 110cm 中国唐苑藏品 刘伟民摄影
Japan Black Pine. Height: 100cm, Width: 110cm. Collector: China Tang Yuan. Photographer: Liu Weimin

14 台湾真柏 *Juniperus chinensis* var. *sargentii* 高 95cm 宽 100cm 中国唐苑藏品 刘伟民摄影
Sargent Savin. Height: 95cm, Width: 100cm. Collector: China Tang Yuan. Photographer: Liu Weimin

15 黑松 *Pinus thunbergii* 高 110cm 宽 120cm 中国唐苑藏品 刘伟民摄影
Japan Black Pine. Height: 110cm, Width: 120cm. Collector: China Tang Yuan. Photographer: Liu Weimin

16 黑松 *Pinus thunbergii* 高 176cm 宽 170cm 中国唐苑藏品 刘伟民摄影
Japan Black Pine. Height: 176cm, Width: 170cm. Collector: China Tang Yuan. Photographer: Liu Weimin

盆景中国 Penjing China

18 2013 "唐苑的世界盆景对话" 国际年度论坛全景报道 报道：CP
Panoramic Reports of The Annual Forum of "the 2013 dialogue to the world Penjing" Reporter: CP

论坛中国 Forum China

30 盆景无国界,思潮汇古都——2013"唐苑的世界盆景对话"国际年度论坛
访谈及图文整理:CP
Penjing Without Borders, Thoughts Converged in the Old Capital—The Annual Forum of "the 2013 Dialogue to the World Penjing" Interview & Reorganizer: CP

中国现场 On-the-Spot

36 真柏的取势造型制作 改作:樊顺利 文:胡光生 地点:西安中国唐苑
The Gesture Determination and Modeling of Sargent Savin Processor: Fan Shunli Author: Hu Guangsheng Place: China Tangyuan in Xi' an city

海外现场 Spot International

42 创作艺术作品之心——真柏盆景铭品"清风"的改作 撰文、制作、供图:【日本】小林国雄
图注:苕源山人
Heart for Creation of Artworks—Recreation of "Breeze", a Famous *Juniperus chinensis* Penjing Author & Artist & Photos Provider: [Japan] Kunio Kobayashi
Pictures: Shaoyuan Recluse

2013"唐苑的世界盆景对话"国际年度论坛
The annual forum of "the 2013 dialogue to the world Penjing"

51 寄情树石 求索真谛 文:王选民
Love in Trees and Stones Exploration for Truth Author: Wang Xuanmin

52 中日盆景交流——对须藤雨伯所提问题的回答 发言人:徐昊 图文整理:CP
China-Japan Penjing communication—Reply the Questions from Mr. Sudo Uhaku Speaker: Xu Hao Reorganizer: CP

54 美国的盆景 发言人:威廉•尼古拉斯•瓦拉瓦尼斯 图文整理:CP
The Origin & History of Penjing in the United States
Speaker: William Nicholas Valavanis Reorganizer: CP

58 西班牙盆景一览 发言人:安东尼奥·帕利亚斯 图文整理:CP
A View of Penjing in Spain Speaker: Antonio Payeras Reorganizer: CP

60 日本盆栽视窗 发言人:小林国雄 图文整理:CP
A Window of Japanese Bonsai Speaker: Kunio Kobayashi Reorganizer: CP

64 论及历史与未来 谈说理念与议题——从盆栽与水石两方面说起 发言人:须藤雨伯
图文整理:CP
Discussed the History and Future Talk about the Concept and Doubts—About

中国盆景赏石
CHINA PENJING & SCHOLAR'S ROCKS
2013 "唐苑的世界盆景对话" 特别专辑
Special Issue of "the 2013 Dialogue to the World Penjing"

Bonsai & Suiseki Speaker: Uhaku Sudo Reorganizer: CP

70 走进宝岛台湾的盆栽艺术 发言人：梁悦美 图文整理：CP
Entering Bonsai Art in Treasure Island, Taiwan Speaker: Amy Liang Reorganizer: CP

75 盆景大小之我见 发言人：杨贵生 图文整理：CP
My Opinion on the Size of Penjing Speaker: Yang Guisheng Reorganizer: CP

76 盆景的精神 发言人：克里斯蒂安·弗内罗；米歇尔·卡尔比昂 图文整理：CP
Spirit of Penjing Speaker: Christian Fournereau & Michèle Corbihan Reorganizer: CP

78 天使的传递 发言人：凯斯图蒂斯·帕特考斯卡斯 图文整理：CP
Angel's Transmission Speaker: Kestutis Ptakauskas Reorganizer: CP

81 印度盆景发展史 发言人：苏杰沙 图文整理：CP
Penjing History in India Speaker: Sujay Shah Reorganizer: CP

84 匈牙利国家盆景概述 发言人：阿提拉·鲍曼 图文整理：CP
Hungarian National Penjing Overview Speaker: Attila Baumann Reorganizer: CP

87 一直在进步—越南盆景 发言人：阮氏皇 图文整理：CP
Keep on Growing-Vietnam Penjing Speaker: Nguyen Thi Hoang Reorganizer: CP

90 马来西亚盆景的起源 发言人：蔡国华 图文整理：CP
The Origin of Malaysia Penjing Speaker: Chua Kok Hwa Reorganizer: CP

96 捷克共和国的盆景进化过程 发言人：斯瓦托普卢克·马特杰卡 图文整理：CP
Process of Penjing Evolution in the Czech Republic Speaker: Svatopluk Matejka Reorganizer: CP

98 至爱盆景 发言人：玛利亚·阿尔博尔莉思·罗斯伯格 图文整理：CP
Loved Penjing Speaker: Maria Arborelius-Rosberg Reorganizer: CP

102 友谊之桥——盆栽 发言人：汤米·尼尔森 图文整理：CP
The Bridge of Friendship— Bonsai Speaker: Tommy Niesen Reorganizer: CP

106 石头的传说 发言人：玛利亚·基亚拉·帕德里齐 图文整理：CP
Tales of stone Speaker: Maria Chiara Padrini Reorganizer: CP

点评 Comments

109 "和谐" 相思 Celtis sinensis 吴成发藏品（见《中国盆景赏石·2012-3》折页）
文：徐昊
"Harmony". Chinese Nettletree. Collector: Ng Shingfat
(Folding Page of the *China Penjing & Scholar's Rocks* 2012-3) Author: Xu Hao

110 榆树 Ulmus pumila 黄经洲藏品（见《中国盆景赏石·2012-8》第 22 页）文：李新
Elm. Collector: Huang Jingzhou. (Page 22 of the *China Penjing & Scholar's Rocks* 2012-8) Author: Li Xin

养护与管理 Conservation and Management

114 树木移植的国际做法（连载二）文：欧永森
The International Method of Tree Transplantation (Serial Ⅱ) Author: Sammy Au

盆艺欣赏 Pot Art Appreciation

116 柯家花园仿古石盆系列欣赏
The Appreciation of the Ke Chengkun's Antique Pot Series

赏石中国 China Scholar's Rocks

117 "一代天骄" 广西三汀石 长 20cm 宽 7cm 高 16cm 魏积泉藏品
"God's Favored One". Guangxi Sanjiang Stone. Length: 20cm, Width: 7cm, Height: 16cm. Collector: Wei Jiquan

118 "塑韵" 百色石 长 14cm 宽 9cm 高 17cm 李正银藏品 苏放摄影
"Sculpture with Special Appeal". Baise Stone. Length: 14cm, Width: 9cm, Height: 17cm. Collector: Li Zhengyin, Photographer: Su Fang

119 "快乐的小精灵" 戈壁玛瑙 长 30cm 宽 27cm 高 39cm 魏积泉藏品
"Happy Elf". Gobi Agate Stone. Length: 30cm, Width: 27cm, Height: 39cm. Collector: Wei Jiquan

120 中国古今名石简谱（连载八）文：文甡
Chinese Famous Rocks Notation(Serial Ⅷ) Author: Wen Shen

123 安徽淮南观赏石（连载二）文：周保友
Ornamental Stone of Huainan Region, Anhui Province (Serial Ⅱ)
Author: Zhou Baoyou

中国唐苑
——一个男人的梦想
China Tang Yuan
—One Man's Dream

文：苏放 Author: Su Fang

Preamble 卷首语

Author Introduction

Su Fang is the president of the China Penjing Artist Association, and initiator, publisher and chief editor of the China Penjing & Scholar's Rocks and the honorary president and plenipotentiary of World Bonsai Stone Culture Association. Besides, he is a contracted musician with Warner Music International Ltd. which is one of the world top three music corporations. Being a major planner, Su participated in the preparations for establishing the state-level China Penjing Artist Association in 1988. He had been secretary-general thereof since 1993 and assuming the post of president since 1999.

> 史称"西有罗马,东有长安"。西安与雅典、罗马、开罗并称为世界四大古都,明洪武二年(1369)废奉元路设西安府,西安即由此而得名。

美国著名歌星鲍勃·迪伦曾经在他的歌里这样唱道:"一个男人要走过多少路,才可以称得上是个男人?"

本月专辑,我们把目光投向了西安的中国唐苑。因为在这里,诞生了世界盆景史上首次"世界盆景论坛"这样的活动。

纵观历史,西安,是成就了很多男人梦想的地方。

西汉初年,刘邦定都关中,取当地长安乡之含意,立名"长安",意即"长治久安"。丝绸之路开通后,长安成为东方文明的中心,史称"西有罗马,东有长安"。西安与雅典、罗马、开罗并称为世界四大古都,明洪武二年(1369)废奉元路设西安府,西安即由此而得名。在中国历史上,有20个王朝政权被认为在西安建都。闻名全球的兵马俑,就在这里发掘面世。1981年联合国教科文组织把西安确定为世界历史名城。

2013年7月当我第一次走访参观建造好后的西安唐苑,独自一人在唐苑里散步的时候,我突然感觉到在这个阔达的庭园园林中,蕴藏着一个男人的梦想。

五年前,有一个这样的男人,他拿出数百万元之巨的人民币,去资助和主办一个开中国盆景展览先河的大型全国盆景展——中国唐风盆景展。就是在那次展览期间,他建造的中国唐苑的名字开始传向全国各地。很多人都在想,一个那么高造价的展览,只能往里面贴钱没有任何盈利,一个那么大的公园,私人花那么多钱来建,从企业财务报表的角度看,也不会有什么太大的利润,这个人为什么要做这样一些事?他图什么呢?

当我把很多人对唐风展、唐苑的疑问抛给这个建造唐苑的男人时,他这样回答我:

"钱是这个世界上最不值钱的东西,文明才是!越是优秀的文明就越值钱!全世界的盆景文明的根源都是从中国的唐代文化文明中传出去的,所以我做的盆景展览的名字就叫'唐风展',我做的这个园林就叫中国唐苑,我认定一件事:盆景是代表了一种中国人的民族文明高度的东西,要发现盆景的本质必须追根寻底,到它发源的土地上来寻找,所以,把唐代传向世界的盆景的文化本质在全世界发扬光大,是我作为一个中国人的义务和责任。这件事上我不关注财务报表上利润栏的数字,我只关心一件事:中国盆景是世界文明中的一种民族文明的代表,一个国家的强大除了军事和经济,还应该有文化的强大,中国人的文化昔日曾经很强大,强大到当年的日本文化中的很多系统的建立和设计都是来自于长安古城,你今天去日本的京都古城看一下当地的建筑就知道中国当年对日本城市建设的影响到底有多么惊人,跟长安古城风格像一个模子刻出来的,所以,唐风展也好,唐苑也好,我其实都是在做一件事,那就是让

Preamble 卷首语

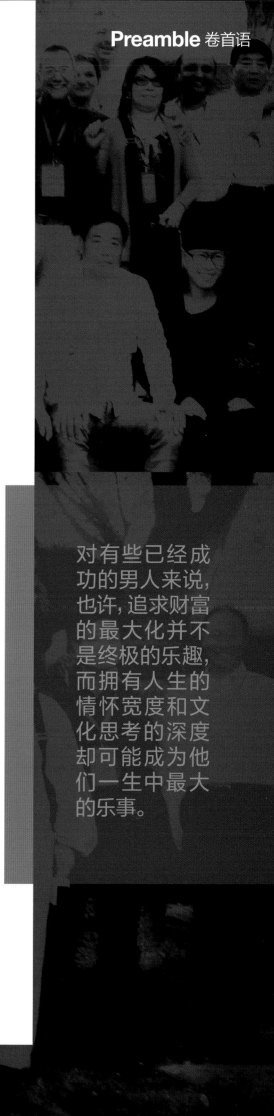

中国盆景文化在全世界发扬光大。"

他的回答让我想起两段话：

1. 被称为"民族精神之父"的18世纪德国哲学家和诗人赫尔德曾经这样说："每一种文明都有自己独特的精神——它的民族精神。这种精神创造一切，理解一切。"

2. 还有一句话这样说：一个失去精神的民族只能任人宰割。

他是一个很低调的人，但我第一次与他相识的时候，就觉得他身上有种很倔强的东西，骨子里的自我和低调的外表融合而出一种大气和坚强，一种在古城的土地上养育出来的一种大国国民的精神和气场。

这次在西安的唐苑举办的"唐苑——2013世界盆景的对话"国际论坛吸引了全球的目光，来到西安的不仅有各国的盆景行业的代表、大师，还有全球最著名的盆景杂志的主编们。很多人对中国能创办这样的让全球盆景人互相交流盆景文化的世界性舞台感到由衷的敬佩。来宾们都异口同声地说：这是一个很棒的创意。中国人，开始挑起下一个时代的盆景文化发展的思考重担了。

短短的3天，在西安唐苑的会议室里，回荡着史无前例的世界盆景交流的声音，而来自日本的盆栽研究家景道家元二世须藤雨伯的振聋发聩的发言令全体各国出席者心灵撼动，他这样说："日本的盆栽是世界盆栽的主流这句话并不是将会永久性地存在下去的。中国盆景的发展将是世界盆栽取得更大发展的巨大原动力，中国盆景应该能改变世界盆栽的未来。中国盆景的历史、思想、宗教、哲学、美学的理论和合理性俱佳，全世界的人易于理解。我相信，中国盆景的发展今后将再一次影响日本盆栽的发展，进而影响世界盆栽的发展。"

对有些已经成功的男人来说，也许，追求财富的最大化并不是终极的乐趣，而拥有人生的情怀宽度和文化思考的深度却可能成为他们一生中最大的乐事。当我在唐苑中散步的时候，我突然想起了英国名著中的鲁滨逊，这样一个人孤独地在荒岛上坚持自己的人生的男人的故事，其实，一个男人在芸芸众生中站起来的过程，往往是一段极其孤独的历程，对外人来说，很多人看到的只是他们成功的外表，但有几个人能知晓他们内心曾经经历的那种骨彻冰心的孤独呢？没有这种孤独历练的人又怎么能懂得一个男人为何对建立"精神文明"这件事儿会如此迷恋呢？

他，一个经常半夜一点起来看盆景并且把已经睡觉的朋友叫醒继续白天的盆景审美和养护讨论的人，就是这样一个对"精神文明"不仅"很迷恋"而且"很投入"的人。

他的名字叫张小斌。

VIEW CHINA
景色中国

黑松 *Pinus thunbergii* 高95cm 宽100cm 中国唐苑藏品 刘伟民摄影
Japan Black Pine. Height: 95cm, Width: 100cm. Collector: China Tang Yuan. Photographer: Liu Weimin

VIEW CHINA
景色中国

黑松 *Pinus thunbergii* 高100cm 宽110cm 中国唐苑藏品 刘伟民摄影
Japan Black Pine. Height: 100cm, Width: 110cm. Collector: China Tang Yuan. Photographer: Liu Weimin

VIEW CHINA
景色中国

台湾真柏 *Juniperus chinensis* var. *sargentii* 高95cm 宽100cm 中国唐苑藏品 刘伟民摄影
Sargent Savin. Height: 95cm, Width: 100cm. Collector: China Tang Yuan. Photographer: Liu Weimin

VIEW CHINA
景色中国

黑松 *Pinus thunbergii* 高110cm 宽120cm 中国唐苑藏品 刘伟民摄影
Japan Black Pine. Height: 110cm, Width: 120cm. Collector: China Tang Yuan. Photographer: Liu Weimin

VIEW CHINA
景色中国

黑松 *Pinus thunbergii* 高176cm 宽170cm 中国唐苑藏品 刘伟民摄影
Japan Black Pine. Height: 176cm, Width: 170cm. Collector: China Tang Yuan. Photographer: Liu Weimin

您能拥有
《中国盆景赏石》
最简便的方法：

成为中国盆景艺术家协会的会员，
免费得到《中国盆景赏石》
如果您是中国盆景艺术家第五届理事会的会员，
《中国盆景赏石》每年赠送您。

成为会员的方法：

1. 填一个入会申请表，把它寄到：
北京朝阳区建外SOHO16号楼1615室 中国盆景艺术家协会秘书处
邮编100022

2. 把会费（会费标准为：每年260元）和每年的挂号邮费（每年12本，共76元）
汇至中国盆景艺术家协会银行账号（见下面）

3. 然后打电话到北京中国盆景艺术家协会秘书处口头办理一下会员的注册登记：
电话是 010-5869 0358

然后……您就可以等着每月邮递员把《中国盆景赏石》给您送上门喽。

中国盆景艺术家协会会费收款银行信息：
开户户名：中国盆景艺术家协会 开户银行：北京银行魏公村支行
账号：200120111017572
邮政地址：北京市朝阳区建外SOHO16号楼1615室
邮编：100022

2013 唐苑的世界盆景对话
DIALOGUE TO THE WORLD PENJING
国际年度论坛全景报道
Panoramic Reports of the International Annual Forum

报道：CP　Reporter: CP

CP 国际年度论坛 The Annual Forum

The Annual Forum 国际年度论坛

Panoramic Reports
of the International Annual Forum

2013"唐苑的世界盆景对话"国际年度论坛嘉宾集体合影
The group picture of the International Annual Forum of the 2013 "Dialogue of World Penjing in Tangyuan"

  2013"唐苑的世界盆景对话"国际年度论坛于2013年10月3至4日在陕西省西安市中国唐苑1号会议室召开并受到了与会中外嘉宾的一致好评,此次国际年度论坛由苏放创意并策划,中国唐苑独家主办。

  2013年10月2日下午,与会中外嘉宾抵达中国唐苑,在唐苑工作人员的引领下,嘉宾们参观游览了唐苑,唐苑内古树成林、松柏挺立,奇石林立、盆景艺术千姿百态,中外嘉宾被唐苑皇家园林般的宏伟风格、气势所折服,不禁发出赞叹。当晚陕西万达实业集团股份有限公司董事长张小斌先生为全体嘉宾安排欢迎晚宴,邀大家在唐苑"本味陕菜"南厅品尝地地道道的陕西菜。中国博大精深的饮食文化给远道而来的外国嘉宾留下了深深的印象。欢迎晚宴上,陕西万达实业集团股份有限公司董事长张小斌先生,中国盆景艺术家协会会长、世界盆景石文化协会名誉会长兼首席全权代表苏放先生,中国盆景艺术家协会副会长、世界盆景石文化协会秘书长樊顺利先生共同为在全世界最新聘请的中国盆景艺术家协会和世界盆景石文化协会国际理事进行了颁证仪式。

  The International Annual Forum of the 2013 "Dialogue of World Penjing in Tangyuan" had been held in No. 1 Conference Room of China Tangyuan in Xi'an City of Shaanxi Province from October 3 to October 4, 2013, and won the unanimous praises from the guests at home and abroad. This international annual forum was designed and planned by Mr. Su Fang and exclusively sponsored by China Tang Yuan.

  On the afternoon of October 2, 2013, guests at home and abroad arrived at China Tang Yuan and visited Tangyuan under the guidance of the staff. And all the guests were attracted by the flourishing forest of ancient trees and rare stones, erect and majestic pines and cypresses as well as the diversities of Penjing art, and could hardly contain themselves from admiring. On the night, Mr. Zhang Xiaobin, the Board Chairman of Shaanxi Wanda Group Co., Ltd. hosted a welcome dinner for all the guests by inviting them to taste the native Shaanxi cuisine in the South Hall of Tangyuan's restaurant called "Flavor of Shaanxi Cuisine". The great and profound Chinese food culture has greatly impressed the foreign guests from afar. At the welcome dinner, Mr. Zhang Xiaobin, the Board Chairman of Shaanxi Wanda Group Co., Ltd., Mr. Su Fang, the President of China Penjing Artists Association concurrent with the Honorary President and Chief Representative of the World Bonsai Stone Culture Association, and Mr. Fan Shunli, the Vice President of China Penjing Artists Association concurrent with the Secretary-general of World Bonsai Stone Culture Association together performed the certificate awarding ceremony for the international council members newly employed for China Penjing Artists Association and World Bonsai Stone Culture Association.

2013 唐苑的世界盆景对话
DIALOGUE TO THE WORLD PENJING

中国盆景艺术家协会会长苏放、陕西万达实业集团股份有限公司董事长张小斌陪同中外嘉宾参观唐苑
Su Fang, the President of China Penjing Artists Association; Zhang Xiaobin, the Board Chairman of Shaanxi Wanda Group Co., Ltd. accompany all the guests on the tour of visiting Tangyuan

陕西万达实业集团股份有限公司副总裁仇胜萍女士在欢迎晚宴上发表讲话
Ms. Qiu Shengping, the Vice President of Shaanxi Wanda Group Co., Ltd. gives speech during the welcome dinner

中国盆景艺术家协会会长、世界盆景石文化协会名誉会长兼首席全权代表苏放先生在欢迎晚宴上发表讲话
Mr.Su Fang, the President of China Penjing Artists Association and the Honorary President and Chief Representative of World Bonsai Stone Culture Association gives speech during the welcome dinner

　　"唐苑的世界盆景对话"国际年度论坛在世界盆景界可谓独树一帜，从来没有一个活动可以如此次国际年度论坛一般，号召到全世界十几个国家的专家、学者相聚一堂，共同探讨世界盆景现今的发展状况，探索世界盆景未来的发展方向。

　　10月3日上午9:00，在一种学术性交流的庄严气氛中，论坛主持人中国盆景艺术家协会副会长、世界盆景石文化协会秘书长樊顺利先生宣布2013"唐苑的世界盆景对话"国际年度论坛开幕！出席此次国际论坛的嘉宾有：陕西万达实业集团股份有限公司董事长张小斌先生，陕西万达实业集团股份有限公司副总裁仇胜萍女士，中国盆景艺术家协会会长、世界盆景石文化协会名誉会长兼首席全权代表苏放先生，陕西唐苑园林观光有限公司韩继明先生，陕西万达实业集团股份有限公司蔡喜萍女士，中国盆景艺术家协会荣誉会长梁悦美女士，中国盆景艺术家协会常务副会长杨贵生先生，中国盆景艺术家协会名誉常务副会长曹志振先生，中国盆景艺术家协会名誉常务副会长李晓先生，中国盆景艺术家协会副会长、世界盆景石文化协会秘书长樊顺利先生，中国盆景艺术家协会副会长芮新华先生，中国盆景艺术家协会副会长申洪良先生中国盆景艺术家协会副会长李城先生，中国盆景艺术家协会会长助理兼中国盆景艺术家协会副秘书长徐昊先生，日本景道家元二世须藤雨伯先生，日本水石协会理事长、春花园BONSAI美术馆馆长小林国雄先生，BCI国际盆栽俱乐部理事（意大利）玛利亚·基亚拉·帕德里齐女士，EBA欧洲盆景协会成员国匈牙利盆景协会副会长阿提拉·鲍曼先生，EBA欧洲盆景协会成员国立陶宛盆景协会会长凯斯图蒂斯·帕特考斯卡斯先生，EBA欧洲盆景协会成员国波兰盆景协会会长沃齐米日·皮特思科先生，EBA欧洲盆景协会成员国瑞典盆景协会代表、瑞典盆景协会理事会成员玛利亚·阿尔博尔莉思·罗斯伯格女士，《国际盆栽》出版人兼主编威廉·尼古拉斯·瓦拉尼斯先生，捷克《盆栽》杂志主编斯瓦托普卢克·马特杰卡先生，西班牙《当代盆栽》主编安东尼奥·帕利亚斯先生，法国《气韵盆栽》杂志出版人克里斯蒂安·弗内罗先生，法国《气韵盆栽》杂志主编米歇尔·卡尔比昂女士，越南盆景协会主席阮氏皇女士，印度盆景大师苏杰沙先生，EBA欧洲盆景协会成员国丹麦盆景协会会长汤米·尼尔森先生，马来西亚盆景雅石协会、BCI国际盆栽俱乐部会员蔡国华先生。

The Annual Forum 国际年度论坛

Panoramic Reports
of the International Annual Forum

陕西万达实业集团股份有限公司董事长张小斌为国际理事颁证
Mr. Zhang Xiaobin, the Board Chairman of Shaanxi Wanda Group Co., Ltd. presents awards to international directors

The International Annual Forum of the "Dialogue of World Penjing in Tangyuan may be appraised as "unique". There has never been an activity as this international annual forum, which has summoned experts and scholars from more than 10 countries all over the world to get together and discuss the current development situation and explore the future development direction of the world Penjing.

At 9:00 am of October 3, in the solemnity of academic communication, the forum host - Mr. Fan Shunli, the Vice President of China Penjing Artists Association concurrent with the Secretary-general of World Bonsai Stone Culture Association announced that the International Annual Forum of the 2013 "Dialogue of World Penjing" in Tangyuan began! The guests attending this international forum are: Mr. Zhang Xiaobin, the Board Chairman of Shaanxi Wanda Group Co., Ltd., Ms. Qiu Shengping, the Vice President of Shaanxi Wanda Group Co., Ltd., Mr. Su Fang, the President of China Penjing Artists Association concurrent with the Honorary President and Chief Representative of the World Bonsai Stone Culture Association, Mr. Han Jiming from Shaanxi Tangyuan Landscape Tourism Co., Ltd., Mr. Cai Xiping from Shaanxi Wanda Group Co., Ms. Liang Yuemei, the Honarary President of China Penjing Artists Association, Mr. Yang Guisheng, the Executive Vice President of China Penjing Artists Association, Mr. Cao Zhizhen and Mr. Li Xiao, the Honorary Executive Vice Presidents of China Penjing Artists Association, Mr. Fan Shunli, the Vice President of China Penjing Artists Association and the Secretary-general of World Bonsai Stone Culture Association, Mr. Rui Xinhua, Mr. Shen Hongliang and Mr. Li Cheng, who are the Vice Presidents of China Penjing Artists Association, Mr. Xu Hao, the President Assistant and Vice Secretary-general of China Penjing Artists Association, Mr. Uhaku Sudo, Japanese Keido Iemoto (headmaster) II, Mr. Kunio Kobayashi, the president of Nippon Suiseki Association and Director of Shunka-en BONSAI Museum, Ms. Maria Chiara Padrini, the council member (Italy) of BCI (Bonsai Clubs International), Mr. Attila Baumann, the Vice President of Penjing Association of Hungary, which is a member country of EBA (European Bonsai Association), Mr. Kestutis Ptakauskas, the President of Penjing Association of Lithuania, which is a member country of EBA (European Bonsai Association), Mr. Wlodzimierz Pietraszko, the President of Penjing Association of Poland, which is a member country of EBA (European Bonsai Association), Ms. Maria Arborelius-Rosberg, the representative and council member of Penjing Association of Sweden, which is a member country of EBA (European Bonsai Association), Mr. William N.Valavanis, the publisher and chief editor of International Bonsai, Mr. Svatopluk Matějka, the editor of Czech magazine Casopis pro milovniky prirody Bonsai, Mr. Antonio Payeras, the chief editor of Spanish magazine Bonsai Actual, Mr. Christian Fournereau, the publisher of French magazine Esprit Bonsaï, Ms. Michèle Corbihan, the chief editor of French magazine Esprit Bonsaï, Ms. Nguyen Thi Hoang, the President of Vietnam Penjing Association, Mr. Sujay Shah, an Indian Penjing Master, Mr. Tommy Nielsen, the President of Penjing Association of Denmark, which is a member country of EBA (European Bonsai Association), and Mr. Dato'Chua Kok Hwa, the member of Malaysia Penjing & Suiseki Society and BCI (Bonsai Clubs International).

极具皇家园林风格的唐苑景色
The beautiful scenary of Tangyuan with royal garden atmosphere

2013 唐苑的世界盆景对话
DIALOGUE TO THE WORLD PENJING

中国盆景艺术家协会会长、世界盆景石文化协会名誉会长兼首席全权代表苏放先生在国际年度论坛上发表讲话
Mr.Su Fang, the President of China Penjing Artists Association and the Honorary President and Chief Representative of World Bonsai Stone Culture Association gives speech during the International Forum

2013"唐苑的世界盆景对话"国际年度论坛主持人：中国盆景艺术家协会副会长、世界盆景石文化协会秘书长樊顺利先生
Mr. Fan Shunli, the Vice President of China Penjing Artists Association and the Secretary-general of World Bonsai Stone Culture Association and the host of 2013 "Dialogue of World Penjing in Tangyuan"

中国盆景艺术家协会会长助理兼中国盆景艺术家协会副秘书长、中国盆景艺术大师徐昊先生
Mr. Xu Hao, the President Assistant and Vice Secretary-general of China Penjing Artists Association and the China Penjing Master

首届西安论坛的议题，试图探讨中外盆景艺术特点以及着眼于世界盆景与各自不同国家的关系、它们又是如何各自影响形成未来的盆景艺术发展趋向等问题。中国唐苑主办首届西安国际论坛的主旨就是，想借此次论坛来向世界宣扬盆栽的发源地——中国的盆景文化、盆景哲学，让国际盆景人体会中国盆景艺术家的艺术哲学，了解拥有悠久盆景历史的中国是从哪个角度对待盆景这一"立体的画，无声的诗"。中国盆景艺术近年来随着中国盆景人越来越多的参与到世界性的盆景活动，以及在本国举办大规模的盆景展会以及开创性的西安国际论坛，吸引到全世界盆景人的瞩目以及亲身参与，中国盆景文化的输出也更加的频繁，中国盆景屹立于世界之巅的那一天已经不远了。

2013"唐苑的世界盆景对话"国际年度论坛现场
2013 "Dialogue of World Penjing in Tangyuan"

嘉宾们聚精会神地听演讲 法国《气韵盆栽》杂志出版人克里斯蒂安·弗内罗先生(左)，马来西亚盆景雅石协会、BCI 国际盆栽俱乐部会员蔡国华先生(右)
Guests attentively listen to the speech Mr. Christian Fournereau, the publisher of French magazine Esprit Bonsaï (left), Mr. Dato'Chua Kok Hwa, the member of Malaysia Penjing & Scholars' Rock Association and BCI (Bonsai Clubs International)(right)

嘉宾们聚精会神地听演讲 EBA 欧洲盆景协会成员国波兰盆景协会会长沃齐米日·皮特思科先生(左)，EBA 欧洲盆景协会成员国立陶宛盆景协会会长凯斯图蒂斯·帕特考斯卡斯先生(右)
Guests attentively listen to the speech Mr. Wlodzimierz Pietraszko, the President of Penjing Association of Poland, which is a member country of EBA (European Bonsai Association)(left), Mr. Kestutis Ptakauskas, the President of Penjing Association of Lithuania, which is a member country of EBA (European Bonsai Association)(right)

The Annual Forum 国际年度论坛

Panoramic Reports
of the International Annual Forum

For the topic of the first Xi'an Forum, attendants have tried to discuss the artistic characteristics of China and foreign Penjing, and focused on the relations between the Penjing and its own country, as well as how they would affect each other and form the development trend of future Penjing art. The first Xi'an International Forum sponsored by China Tang Yuan is intended to promote the Penjing culture and philosophy of China – the birthplace of bonsai to the world by this forum, providing the international Penjing artificers with an opportunity to feel the artistic philosophy of China Penjing artists and understand how we appreciate Penjing – "a stereo painting and silent poem" in China, a country with the profound Penjing history. With Chinese Penjing artificers, the Chinese Penjing art has more and more participated in the world Penjing activities in recent years. The grand

唐苑内千姿百态的盆景
Various Penjing in Tangyuan

Penjing exhibitions and creative Xi'an International Forum held in China have attracted the eyes of all Penjing artificers over the world. And the Chinese Penjing culture has been exported more and more frequently. The day for the China Penjing to top the world is not far away.

中国盆景艺术家协会名誉会长梁悦美女士
Ms. Liang Yuemei, the Honarary President of China Penjing Artists Association

中国盆景艺术家协会常务副会长杨贵生先生
Mr. Yang Guisheng, the Executive Vice President of China Penjing Artists Association

日本水石协会理事长、春花园BONSAI美术馆馆长小林国雄先生
Mr. Kunio Kobayashi, the President of Nippon Suiseki Association and Director of Shunka-en BONSAI Museum

百米瀑布
Amazing Waterfall

2013 唐苑的世界盆景对话
DIALOGUE TO THE WORLD PENJING

中外嘉宾在唐苑合影留念
The Chinese and foreign guests take photos in Tangyuan

嘉宾们聚精会神地听演讲 印度盆景大师苏杰沙先生（左）、越南盆景协会主席阮氏皇女士（右）
Guests attentively listen to the speech Mr. Sujay Shah, an Indian Penjing Master(left), Ms. Nguyen Thi Hoang, the President of Vietnam Penjing Association(right)

日本景道家元二世须藤雨伯先生
Mr. Uhaku Sudo, Japanese Keido Iemoto (headmaster) II

《国际盆栽》出版人兼主编威廉·尼古拉斯·瓦拉瓦尼斯先生
Mr. William N.Valavanis, the publisher and chief editor of International Bonsai

BCI 国际盆栽俱乐部理事（意大利）玛利亚·基亚拉·帕德里齐女士
Ms. Maria Chiara Padrini, the council member (Italy) of BCI (Bonsai Clubs International)

The Annual Forum 国际年度论坛

西班牙《当代盆栽》主编安东尼奥·帕利亚斯先生
Mr. Antonio Payeras, the chief editor of Spanish magazine Bonsai Actual

法国《气韵盆栽》杂志出版人克里斯蒂安·弗内罗先生、法国《气韵盆栽》杂志主编米歇尔·卡尔比昂女士
Mr. Christian Fournereau, the publisher of French magazine Esprit Bonsaï, Ms. Michèle Corbihan, the chief editor of French magazine Esprit Bonsa

EBA 欧洲盆景协会成员国匈牙利盆景协会副会长阿提拉·鲍曼先生
Mr. Attila Baumann, the Vice President of Penjing Association of Hungary, which is a member country of EBA (European Bonsai Association)

EBA 欧洲盆景协会成员国立陶宛盆景协会会长凯斯图蒂斯·帕特考斯卡斯先生
Mr. Kestutis Ptakauskas, the President of Penjing Association of Lithuania, which is a member country of EBA (European Bonsai Association)

EBA 欧洲盆景协会成员国瑞典盆景协会代表玛利亚·阿尔博尔莉思·罗斯伯格
Ms. Maria Arborelius-Rosberg, the representative and council member of Penjing Association of Sweden, which is a member country of EBA (European Bonsai Association)

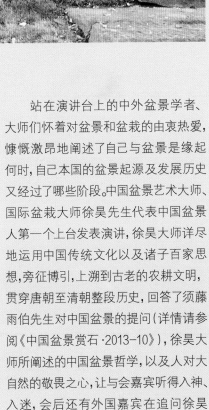

站在演讲台上的中外盆景学者、大师们怀着对盆景和盆栽的由衷热爱，慷慨激昂地阐述了自己与盆景是缘起何时，自己本国的盆景起源及发展历史又经过了哪些阶段。中国盆景艺术大师、国际盆栽大师徐昊先生代表中国盆景人第一个上台发表演讲，徐昊大师详尽地运用中国传统文化以及诸子百家思想，旁征博引，上溯到古老的农耕文明，贯穿唐朝至清朝整段历史，回答了须藤雨伯先生对中国盆景的提问（详情请参阅《中国盆景赏石·2013-10》），徐昊大师所阐述的中国盆景哲学，以及人对大自然的敬畏之心，让与会嘉宾听得入神、入迷，会后还有外国嘉宾在追问徐昊大师关于盆景哲学——"意境"的定义。

嘉宾们聚精会神地听演讲
EBA 欧洲盆景协会成员国丹麦盆景协会会长汤米·尼尔森先生
Guests attentively listen to the speech Mr. Tommy Nielsen, the President of Penjing Association of Denmark, which is a member country of EBA (European Bonsai Association)

2013 唐苑的世界盆景对话
DIALOGUE TO THE WORLD PENJING

让人流连忘返的唐苑景色
Fantastic scenery in Tangyuan

With their true love for Penjing, the Chinese and foreign scholars and masters at the podium have made impassioned speeches about how they fell in love with Penjing, how Penjing in their country originated and the development steps. Mr. Xu Hao, the Chinese Penjing art master and the international bonsai master, was the first to speak on the stage. Master Xu Hao has answered Mr. Uhaku Sudo's question by applying the Chinese traditional culture and thoughts of the ancient philosophers, quoting copiously from the ancient cultivation culture and throughout the whole history from the Tang Dynasty to the Qing Dynasty. (See China Penjing & Scholars' Rock 2013-10 for details.) The attendants have been entranced and fascinated by the Chinese Penjing philosophy and people's admiration and reverence towards nature described in Master Xu Hao's speech. After the meeting, the foreign guests were still asking Master Xu Hao about the definition of Penjing philosophy – "artistic conception".

越南盆景协会主席阮氏皇女士
Ms. Nguyen Thi Hoang, the President of Vietnam Penjing Association

捷克《盆栽》杂志主编斯瓦托普卢克·马特杰卡先生
Mr. Svatopluk Matějka, the editor of Czech magazine Casopis pro milovniky prirody Bonsai

中外嘉宾参观唐苑的秀丽景色
All the guests enjoy the splendid landscape in Tangyuan

The Annual Forum 国际年度论坛

Panoramic Reports
of the International Annual Forum

EBA 欧洲盆景协会成员国瑞典盆景协会代表、瑞典盆景协会理事会成员玛利亚·阿尔博尔莉思·罗斯伯格女士(左)，捷克《盆栽》杂志主编斯瓦托普卢克·马特杰卡先生(右)
Ms. Maria Arborelius-Rosberg, the representative and council member of Penjing Association of Sweden, which is a member country of EBA (European Bonsai Association)(left), Mr. Svatopluk Matějka, the editor of Czech magazine Casopis pro milovniky prirody Bonsai(right)

法国《气韵盆栽》杂志主编 米歇尔·卡尔比昂女士想记录下这精彩的瞬间
Ms. Michèle Corbihan, the chief editor of French magazine Esprit Bonsaï wants to record this wonderful moment

印度盆景大师苏杰沙先生
Mr. Sujay Shah, an Indian Penjing Master

马来西亚盆景雅石协会、BCI 国际盆栽俱乐部会员蔡国华先生
Mr. Dato'Chua Kok Hwa, the member of Malaysia Penjing & Scholars' Rock Association and BCI (Bonsai Clubs International).

EBA 欧洲盆景协会成员国丹麦盆景协会会长汤米·尼尔森先生
Mr. Tommy Nielsen, the President of Penjing Association of Denmark, which is a member country of EBA (European Bonsai Association)

中国盆景艺术家协会会长、世界盆景石文化协会名誉会长兼首席全权代表苏放先生向外国嘉宾介绍《中国盆景赏石》
Mr. Su Fang, the President of China Penjing Artists Association and the Honorary President and Chief Representative of World Bonsai Stone Culture Association introduces China Penjing & Scholar's Rocks to foreign guests

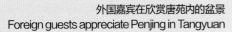

外国嘉宾在欣赏唐苑内的盆景
Foreign guests appreciate Penjing in Tangyuan

唐苑的世界盆景对话
DIALOGUE TO THE WORLD PENJING

中国盆景艺术家协会和世界盆景石文化协会国际理事颁证仪式
The award ceremony of international director of China Penjing Artists Association and World Bonsai Stone Culture Association

日本景道家元二世须藤雨伯先生,日本水石协会理事长、春花园BONSAI美术馆馆长小林国雄先生也先后上台阐述自己对于盆景深入的研究从而得出一种具有日本文化特色的盆景理念。西方盆景开始的历史不很久远,但是每个国家都有各自的创新之处无论是造型还是技术方面,这些都让与会的中国嘉宾看到了盆景在另一个地域范围发展的现状,这一点也会潜移默化地促使中国盆景今后的发展方向更加宽广,视野更加开阔。

2013年10月5日,为期两天的2013"唐苑的世界盆景对话"国际年度论坛圆满结束,所有嘉宾都表示两天时间虽然不长,但是短短两天却可以纵观当代中外盆景历史、现状和未来趋向,国际论坛办得简直是太充盈了!

当晚,在位于唐苑的"本味陕菜"餐厅为第二天就要返程的与会中外嘉宾举行了别致的欢送晚宴,唐苑之行在外国嘉宾的脑海里刻上了深深的烙印,这里的盆景、这里的美食浓缩着人生中两大追求——与自然的亲近,对食物的追求。

Mr. Uhaku Sudo, Japanese Keido Iemoto (headmaster) II, and Mr. Kunio Kobayashi, the president of Nippon Suiseki Association concurrent with the Director of Shunka-en BONSAI Museum, have successively stated their further research on Penjing and the Penjing concept with Japanese culture characteristics they obtained from the research. The history of western Penjing has not been long, but every country has its creative part in no mater modeling or technique, and that has shown the current development of Penjing in another territorial scope to Chinese guests, and gradually and unconsciously widens the future development direction and view of China Penjing.

On October 5, 2013, the two-day International Annual Forum of the 2013 "Dialogue of World Penjing in Tangyuan successfully ended. All the guests have expressed that, during the short time of two days, they have taken a panoramic view for the history, current situation and future trend of the modern Chinese and foreign Penjing. The international forum was amazingly informative!

On the night, an unconventional farewell dinner was hosted in Tangyuan's restaurant called the "Flavor of Shaanxi Cuisine" for all the guests who would leave on the next day. The tour of Tangyuan has left the foreign guests with an indelible impression. The Penjing and food here has concentrated the two pursuits in one's life – pursuits for the nature and the food.

The annual forum 国际年度论坛

Panoramic Reports
of the International Annual Forum

2013"唐苑的世界盆景对话"国际年度论坛是世界盆景活动的一个创新,这个创新会掀起世界盆景界的一个新风尚——齐聚中外盆景人,探讨中外盆景,为世界盆景的发展尽一份力!

(左起)日本水石协会理事长、春花园BONSAI美术馆馆长小林国雄先生,陕西万达实业集团股份有限公司董事长张小斌先生,中国盆景艺术家协会会长、世界盆景石文化协会名誉会长兼首席全权代表苏放先生,中国盆景艺术家协会副会长李城先生,中国盆景艺术家协会常务副会长杨贵生先生
(from left) Mr. Kunio Kobayashi, the president of Nippon Suiseki Association and Director of Shunka-en BONSAI Museum; Mr. Zhang Xiaobin, the Board Chairman of Shanxi Wanda Group Co., Ltd.; Mr. Su Fang, the President of China Penjing Artists Association concurrent with the Honorary President and Chief Representative of the World Bonsai Stone Culture Association; Mr. Li Cheng, who are the Vice Presidents of China Penjing Artists Association; Mr. Yang Guisheng, the Executive Vice President of China Penjing Artists Association

The International Annual Forum of the 2013 "Dialogue of World Penjing in Tangyuan is a creation of the world Penjing activities. And this creation will lead a new tendency for the world Penjing – gathering Penjing artificers at home and abroad to discuss the world Penjing and make contribution to its development!

欢送晚宴上,外国嘉宾忘情起舞歌唱
International guests sing and dance during the farewell dinner

2013 唐苑的世界盆景对话
DIALOGUE TO THE WORLD PENJING

盆景无国界，思潮汇古都
——2013 "唐苑的世界盆景对话" 国际年度论坛
Penjing Without Borders, Thoughts Converged in the Old Capital
—The Annual Forum of "the 2013 dialogue to the world Penjing"

访谈及图文整理：CP　Interview & Reorganizer: CP

【美国】威廉·尼古拉斯·瓦拉瓦尼斯
[America] Valavanis Willima Nicholas
《国际盆栽》杂志出版人兼主编
Publisher and Editor for *International BONSAI* Magazine

西安唐苑举办的2013世界盆景对话论坛对我而言非常有价值，尤其作为一名研究盆景的学者和盆景杂志出版人来说。看到如此多来自不同国家的人们聚在一起，分享每一个值得尊敬的国家的盆景人讲述有关盆景的知识，令我很惊讶。我很高兴被邀请到这个活动中并且希望我的陈述可以被大家所理解，同时祝贺此次论坛的成功。

我认为2013 "唐苑的世界盆景对话" 国际年度论坛对世界盆景和盆栽界是一次有价值的贡献。我认为过去没有这样的论坛会，这么多国家的人被聚在一起分享有关盆景的知识。我只是希望其他有盆景历史的主要国家也参与进来，如英国、澳大利亚和德国，以便我能了解更多相关的信息以及这些国家举办的盆景活动。我惊叹于对整个世界的广大盆景爱好者有如此洞察力的有组织的人们，他们有那么伟大的想法而且能胜利的完成。当然，这是一个不朽的功绩。

The 2013 Dialogue to the World Penjing was quite valuable to me, especially as a bonsai student and magazine publisher. It was amazing to see so many people from different countries together sharing information on each of their respected countries. I was honored to be invited to this event and hope my presentation was understood by the others and perhaps helped to the forum's success.

I think the forum was a valuable contribution to the Penjing and bonsai community around the world. I don't think there was ever such a diverse group of people gathered sharing information. I only wish other major countries where bonsai has an established history, such as UK, Australia and Germany were represented, so I could learn more about the introduction and activities in those countries. It was truly amazing that the organizing person with such an insight to the entire world wide bonsai community had such a grand idea, and was able to pull it off as well. Surely, it was a monumental feat.

【日本】小林国雄
[Japan] Kunio Kobayashi
日本水石协会理事长
日本春花园BONSAI美术馆馆长
the President of Nippon Suiseki Association and Curator of Shunkaen BONSAI Museum

对于西安论坛，不同国家对盆栽的价值观和思考方式也是不同的，因此，思考盆景的未来是极为重要的事情。希望可以讨论与盆栽相关的美学、哲学、唯心论、死生论和具有生命的盆景艺术的理念等。另外，期待展场和论坛会场可以在同一场所。

For the Xi'an Forum, the values and ways of thinking for bonsai are different in each country. Therefore, thinking about the future of Penjing is extremely important. I hope we can discuss the relevant aesthetics, philosophy, idealism, life and death theory, the concept of Penjing arts with life and so on. Additionally, I look forward to that the exhibition hall and the forum venue can be at the same place.

西安論壇について、国によって盆栽に対する価値観や考え方が違うので将来のことを考えると重要な事である。盆栽における美学・哲学・精神論・死生観や生命ある盆栽芸術の理念などを討論したい。展示会場と会議の場所が同じことが望ましかったであろう。

Forum China 论坛中国

【日本】须藤雨伯
[Japan] Uhaku Sudo

景道家元二世
Keido Iemoto (headmaster) II

Xi'an Forum is an extraordinarily great planning. It means a lot for the bonsai practitioners all over the world to gather on the land of China and participate the meeting in Xi'an. The meaning is really profound for people representing the world bonsai culture to gather on the land of historical origin of bonsai, and present the concept or current situation of bonsai with their own national characteristics. Meetings of this kind shall be held in the future and I hope the forums will be with thematic characteristics.

I think that the future bonsai culture will become the savior-like culture for the human in the world. Especially that people can understand the definition and concept of Penjing through bonsai. And for the future of human and especially for the developed countries, there may be spiritual guidance for "health", "non-aging" and "immortality". Penjings, in my definition, are not limited in the expression of beauty and artistry, but as a culture to deliver health for people on the earth. And this is contained in the original meaning of Penjing. Therefore, I hope China, with a Penjing history of 3000 years, can promote the world to learn the theme of health via Penjing and bonsai as the organizer.

西安的论坛是极为优异的企划。世界的盆栽人在中国的土地上、在西安参会的意义非凡。代表世界盆栽文化的人们聚集在作为盆栽历史源头的土地上，发表具有各自国家特征的盆栽理念或盆栽现状，着实是意义深刻。这样的会议今后一定要进行下去，希望今后开展有主题特色的论坛。

我认为今后的盆栽文化将成为世界人类文化的"救世主"。人们通过盆栽来理解盆景的定义与理念，可能会对"健康"、"不老"、"不死"等精神具有指导性作用，这对于人类的未来，特别是对于发达国家来说极为重要。盆景不应局限于呈现美感和展现艺术性，更应作为一种传递健康的文化（盆景本身即具有这层意义）来发扬光大。因此，我希望拥有3000年盆景历史的中国作为论坛主办方，通过盆景、盆栽（的交流活动）促进全世界来关注健康方面的主题。

西安おいての論壇は大変素晴らしい企画であったと感じた。世界の盆栽人が中国の地においてしかも西安で行われたことは大変意義のあったことと思う。盆栽歴史の源流となる地で世界の盆栽文化を代表する人たちが集まり、それぞれの国の特徴ある盆栽理念・現況を発表し合うことは、大変意義深いものであった。今後は、テーマを絞って特徴ある論壇にすることを希望いたす。このような会議は今後も是非行ってほしいと思う。

私はこれからの盆栽文化は世界人類の救世主になれる文化と考える。特に盆景の定義・理念を世界の人々に盆栽を通じて理解することにより、人類の未来に特に先進国に「元気」「不老、不死」の精神性を導くことが出来ると考える。私の考える盆景は美の表現や芸術性に留まらず、盆景本来の意味する元気を世界の人たちに発信する文化として広めていきたいと考える。そのためにも主催は中国であり、盆景3000年の歴史に培われた盆栽・盆景により世界に元気をテーマに勉強を進めて頂きたいと思う。

【瑞典】玛利亚·阿尔博尔莉思·罗斯伯格
[Sweden] Maria Arborelius-Rosberg

EBA（欧洲盆景协会）成员国瑞典盆景协会代表
the representative of Swedish Bonsai Association

The garden was so big and beautiful, very impressive! So fantastic with waterfalls! It was very nice to take a walk by myself in the after-noon. The accommodation was also very grand! All the restaurants were very nice with good service.

The annual forum was very well organized. Maybe some of the presentations were too long so it would have been good if every-one had information before to keep it short. It is nice to hear what the Chinese Penjing people talked about and nice to meet and talk with people from all over the world. I feel very lucky to have made so many new friends!

See above. Also I liked to see how close we are to each other with this nice art form. Better understanding of the differences between Penjing and Bonsai. To realize that all comes from China and a good inspiration to me! It makes me want to learn more!

西安唐苑成功地举办了此次年度国际论坛。西安唐苑面积很大，环境优美，令人印象深刻! 令人惊奇的瀑布景观! 下午，一个人在园中惬意的散步。唐苑酒店非常的豪华，所有餐厅装潢得都很漂亮，服务很周到。

年度国际论坛组织得井井有条。也许有的嘉宾的演讲太长，如果每个人能提前控制时间就更好了。非常有幸能听到中国盆景人谈论盆景。很高兴能与各国盆景人见面畅谈。能交到这么多的新朋友，我很幸运。

除了以上说到的事情之外。我非常高兴地感觉到，通过这种艺术形式大家变得更紧密起来。更好地理解了盆景与盆栽的区别。意识到盆景文化源起于中国，这给了我莫大的启示，激发了我想学到更多关于盆景文化的热情。

【匈牙利】阿提拉·鲍曼
[Hungary] Attila Baumann
EBA（欧洲盆景协会）成员国匈牙利盆景协会副会长
the Vice President of Hungarian Bonsai Association

西安的国际论坛给所有国外友人留下了难忘的回忆。在西安举办的国际论坛，是一个非常好的创意，可以把各个国家的代表聚到一起讨论当下的世界盆景文化，设想盆景文化的未来。首届的国际论坛为不同国家的盆景文化提供了一个很好的相互交流的机会，国家间可以相互沟通理解不一样的盆景文化。通过此次在西安唐苑举办的国际论坛，我们找到了彼此身上的共同点——热爱盆景。我从中了解到许多新的角度，怎样从不同角度审视盆景文化和试着体会他人对盆景文化的理解。通过这次活动，人们相互间建立了新的友谊，通过深入探讨，我们开始了解盆景和盆栽的不同。作为匈牙利盆景协会的代表，我非常荣幸可以在台上与大家分享匈牙利盆景的历史和发展轨迹虽然匈牙利国土面积不大，但是有很多人都致力于把盆景文化传播到全国甚至全世界。通过沟通交流，交换不同意见和盆景技法以及在盆景制作方面的成功经验，我们达到了相互学习的目的。

西安论坛的举办地是著名的私家园林西安唐苑，整个唐苑的构造证明了中国悠久的（举世闻名的）园林文化。极具观赏性的树木以及赏石给人们带来了别样的感觉。大型的树木看起来十分自然，人们可以在其间放空思想、尽情放松。整个唐苑的规划布局给人们提供了巨大的美的享受。夜晚的瀑布以及五彩缤纷的灯光效果又让我们感受到一种新的别样风格。唐苑的地貌多为陡坡宛如置身于大自然，很难想象在十几年间就可以建成拥有7个用餐区的唐苑。

我非常喜欢在西安唐苑举办的年度国际论坛，这种国际论坛的形式很少见，不仅邀请到了中国的周边国家，世界上其他欧洲国家也在邀请之列。非常有幸可以了解到对盆景文化的不同看法，还有中国人民对盆景文化的看法。

The Annual Forum held in Xi'an gave me a great impression. This event is a very good start to get many countries together and discuss the current situation of the Penjing culture in the world, and to make ideas about the future. This first forum provided a good opportunity to learn the beginnings of this culture in different countries, and understand the different approaches and feelings regarding the bonsai. With this we made the first step to find common things among the people, who likes bonsai. I learned here many new aspect, how to see this culture from different perspectives, and the thinking of individual people. New beautiful friendships were established during this event, and deep discussions were started to understand bonsai and Penjing.

For Hungary it was a very big pleasure to present our history and the present development stage and show the participants, that our country is small, but there is a high engagement to spread this culture in the country and also in the world.

We can learn from each other by exchanging different views on the techniques and share the achievements made by all the countries.

Xi'an Tangyuan where is the Annual Forum being held. The engagement to create such a garden is to be appreciated, the built garden proves the high garden culture in China (which is already well known), and the combination of trees stones and windmill stones creates additional emotions in the people. Because of the big transplanted trees it looks natural, where people can relax and free up the mind. The garden offers more than one day entertainment for the visitors with its beautiful arrangements. The waterfall as a special feature of this garden allows you to get new impressions in the night with its colorful lighting solutions. Also the hilly areas looked very natural; it is nearly an impossible achievement to build this garden with its 7 restaurants only in 10 years.

I liked very much the Annual Forum, because it was unique in the form, that not only neighbored countries were invited but other foreign countries from all over the world. I appreciate the opportunity to understand the different thinking about the bonsai culture, and to understand the Chinese thinking about it.

【意大利】玛利亚·基亚拉·帕德里奇
[Italy] Maria Chiara Padrini
BCI（国际盆景俱乐部）理事
BCI Director

在西安举办的国际论坛是一个我十分欣赏的主意，把来自全世界并且在各自国家拥有重要背景的人士齐聚在一起，向他们介绍中国盆景，与此同时互相交流建立联系。当我还是意大利盆栽水石大师协会主席时，我在欧洲也试图做了相同的事，把不同国家的艺术家聚到一起，我相信他们的制作表演对于现在有技术的人来说没有很大的意义，也许教育意义也不会很大，因为现在盆景技艺的发展已经很成熟，很短时间内就可以对树木进行很大的改造，但是，从特定意义上来说，交流经验和文化却变得越来越重要。

西安论坛这种实践非常值得全球盆景人学习。我十分赞赏这次的国际论坛活动，而且我认为应该越来越多的举办这种活动。通常情况下，在大会上经常进行现场制作表演，但是这种形式从技术角度来讲受限很大。通常情况下，在我们的国家，我们都是从传闻中了解到其他国家的文化和美学而不是通过自己的亲身经历，这种情况下就会造成误解和错误的评价。盆景制作表演还是会对刚入门的盆景新手很有用处，但是对于更有经验的盆景人来说，这些人需要更广阔的空间交流想法和经验。用语言这种形式，一种有力的方式去克服不同文化间的理解困难，并且演变成一种真正的友谊和兄弟之情。

Forum China 论坛中国

I quite liked the idea of putting together people from all over the world with important background in their respective countries to let them know the Chinese Penjing and, at the same time, to share information and to establish contacts. When I was President of the Italian National Association of Bonsai and Suiseki instructors, I tried to create something similar in Europe with artists from various countries. I believe the demonstrations in themselves have little meaning now for skilled individuals, and perhaps a little education because heavy works are imposed on trees in a very short time, but it becomes important to exchange experiences and culture on the specific.

I really appreciated the Forum and I think we should give more time and space for initiatives like this. Typically in the congresses are held demonstrations but this situation is largely limited to technical aspects. Often in our countries we know different culture and aesthetics by hearsay rather than direct experience, which can foster misunderstandings and errors of assessment. While the conference with demonstrations can remain useful and instructive for those who are starting to practice, it becomes necessary to a more experienced people to give space for exchanging ideas and experiences. The conversational aspect also represents a formidable means to overcome the difficulties of understanding between different cultures and becomes real means of brotherhood and friendship.

【法国】克里斯蒂安·弗内罗
[France] Christian Fournereau
法国《气韵盆栽》杂志出版人
Chief Publication for Esprit Bonsa Magazine

【法国】米歇尔·卡尔比昂
[France] Michèle Corbihan
法国《气韵盆栽》杂志主编
Chief Redactor for Esprit Bonsa Magazine

国际论坛活动这是一个非常有意义的活动，此次国际论坛的水平相当高。对于我们"外国人"而言，中国大师所做的贡献以及他们的盆景"哲学"特别有意义，对所有人来说都是一次新鲜的体验。日本大师也对此抱有极大的兴趣。

我们对欧洲盆景有一个良好的认识，但是我们对世界其他国家的盆景却一无所知。这的确是一个很好的能与世界上所有其他国家分享盆景知识的起点。通过运用一个重要的哲学维度，我们对亚洲地区的盆景艺术有了更为客观的认识。这同样也是一个很好的机会来与其他国家的人们进行沟通和交流，这也将有助于两国人民更好地理解彼此的盆景艺术和文化。

国际论坛的举办地西安唐苑不但物种丰富而且非常有趣。我们必须在2小时内完成参观，但我认为需要一天时间来通过不同的观点来欣赏花园中各种各样的植物。同样值得一提的就是我们有幸欣赏了收藏的茶壶。同样，这个花园还有壮丽的风景值得欣赏。总而言之，这个花园真的是太华丽了。

进行国际论坛的想法真的非常有趣。由于是第一次参加盆景展会，众多国际化盆景潮流的代表们都聚集在一起交流各自的观点。可以通过分享知识和实践经验来发展全球各地的盆景艺术。盆景爱好者的一大兴趣在于了解全球各地的盆景到底发生了些什么。因此能够与走在盆景界潮流尖端的人们沟通和交流是一件多么好的事啊。盆景展会参与者之间的演讲非常不同但是却非常有趣。中国、日本和亚洲的其他国家将大量的技术和哲学知识传播到世界上其他国家，相反，世界上其他国家也有许多值得传播的技术信息，这些信息也值得我们相互学习。

The Annual Forum of "The 2013 dialogue to the world Penjing" held by Xi'an Tang Yuan was a very interesting undertaking and the contributions were partially from high level. For our "foreign ears", the contribution of the Chinese masters and their vision of the Penjing "philosophy" were particularly interesting, and we think it could be from great interest for all of us to renew this experience. The Japanese masters were of great interest too. If have a good knowledge of the European Penjing we hardly know anything about the Penjing in the rest of the world. It was indeed an excellent initiative to share the knowledge of Penjing with all the other countries in the world. That I think will enable us to bring all of this in perspective not only in our country but also around us. We have acquired a more objective vision of the art of Penjing in Asia with an important philosophical dimension to it.

It was also a good opportunity to meet and reach out people from other countries to initiate exchange that will help peoples and culture to better understand each other and to move around for the art of Penjing.

Xi'an Tang Yuan has a great richness and his very interesting. We had to visit it within 2 hours, but I think an entire day would be necessary to appreciate the different points of view and the great variety of plants that it offers. It was also very interesting to get the chance to admire the superb collection of teapots. This garden has magnificent landscapes that what one can admire all along. It was just magnificent.

The idea of an international forum is really interesting. For the first time in the story of the Penjing, the representatives of numerous international trends of the Penjing could seat together to exchange their views. Sharing knowledge and best practices could be an opportunity to develop the art of Penjing throughout the world. Penjing enthusiasts take a great interest in what's going on around the world concerning Penjing. It is therefore a good thing to be able to meet those who perform in the field of Penjing. The speeches exchanged between those participating were very different and interesting. Driven by China and Japan, Asia has a lot to transmit to the rest of the world in terms of techniques and philosophy. But the rest of world has also a lot to transmit; we all have a lot to learn to each other.

【西班牙】安东尼奥·帕利亚斯
《当代盆栽》杂志执行主编
[Spain] Antonio Payeras
Editor and Director for *Bonsai ACTUAL* Magazine

这里举办的2013世界盆景对话是一个探索了解盆景的新理念的极好的机会。这个论坛云集了不同的国家且促进我们相互间的了解。与此同时，这个论坛帮助我们逐渐理解由于各国盆景的不同理念产生的差异。我认为，这样的国际论坛好似一座桥梁连接起不同的国家，提供了一个以不同方式评价盆景的更好的交流平台。

The Annual Forum of "the 2013 Dialogue to the World Penjing" was an excellent opportunity to discover new ways of understanding the bonsai. To find out what makes various countries and has helped us to mutual understanding. In the same way has reduced understandings between our different views of bonsai.

【越南】阮氏皇
越南盆景协会主席
[Vietnam] Nguyen Thi Hoang
the President of Vietnam Bonsai Association

在西安举办的国际论坛是一个非常不错的想法，它可以提供一个很好的机会来聚集全球各地的盆景爱好者。我很高兴能参加这个论坛，我也感到很荣幸能受到邀请。

我非常喜欢西安国际论坛。它将全球各地的盆景爱好者聚集在一起。举行论坛期间，盆景爱好者们也建立了深厚的友谊。

此外，人们也有机会像其他所有盆景爱好者介绍他们所在国家的盆栽和盆景。因此，我们也有机会了解了更多关于不同国家的对于盆景概念。对我们来说，通过这次机会（所有盆景爱好者加入论坛），我们可以拉近彼此之间的距离，有更多的了解，并成为真正的朋友。

The International Forum is a great idea which offers a good chance to gather the Penjing lovers all around the world. I am so happy to attend the forum, and I feel honored to be invited.

I really like the international forum. It connects Penjing lovers around the world. During the forum, Penjing friendship is set up. Moreover, people have the chance to give the introduction about bonsai and Penjing in their countries to all Penjing lovers. Thus, we understand more about different Penjing concept of different countries. It is also a good chance for us (all Penjing lovers who joined the forum) to be near to each other, and then understand more, and become true friends.

【印度】苏杰沙
印度盆景大师
[India] Sujay Shah
India Penjing Master

在西安唐苑举办的年度论坛，看得出它是经过精心仔细的筹备。它非常有意思并且有大量的深度信息。这是一场坦诚又实事求是的讨论，演讲者们自由的发挥！总的来说，令我受益匪浅！

我喜欢2013"唐苑的世界盆景对话"国际年度论坛这项活动，在这个论坛上我们是主角，我们演讲，观看PPT演示并就不同的议题展开讨论。这是一个非常好的学习机会。最大的收获是我们都很受鼓舞，激励我们为盆景事业而工作并传播这项艺术。

Annual Forum at Xi'an Tang Yuan was very meticulously planned. It was very interesting and had a lot of depth of information. The discussion was frank and realistic and speakers were given liberty and freedom! Overall a lot of learning!

We like the 2013 Dialogue in which we spoke, saw the Power Point and discussed. It was a very good learning lesson. The biggest harvest was that we were very inspired to work and spread this art.

【马来西亚】蔡国华
国际盆栽俱乐部会员 马来西亚盆景雅石协会会员
[Malaysia] Dato'Chua Kok Hwa
Member of Bonsai Clubs International, Member of Malaysia Bonsai N Stone Association

西安举办的2013"唐苑的世界盆景对话"国际年度论坛让我了解到各国有关盆景的历史、树种、造型、风格等方面的知识，提高了我对盆景的审美能力。

我很享受在西安举办的2013"唐苑的世界盆景对话"国际年度论坛。此次论坛的意义在于我们可以学习到其他国家关于盆景的理念、哲学思想和经验方面的知识，以及各国人如何提升盆景的某些观点和见解。本次论坛时间紧凑，但是最后进展得非常顺利。日本小林国雄先生呈现的2幅画让我们很感兴趣，我们思考这两幅画的区别。一幅画画着栽种在盆里的花，象征花开的美丽令我们惊叹，另一幅画画着栽种在盆中的盆景象征着盆景的美丽表达着人类的思想。

I feel "the Annual Forum of the 2013 Dialogue to the World Penjing" in Xi'an allowed us to learn about Penjing in other countries in terms of history, species, form, style, etc. as well as views and perspectives on improving Penjing.

I enjoyed the Annual Forum of the 2013 "Dialogue to the World Penjing" in Xi'an. The significance of this dialogue was to learn from others about their views, thoughts and experiences on Penjing. The limitation of time was a main setback for this dialogue but in the end it all went well. The drawing of a flower in a pot and a Penjing in a pot by Kunio Kobayashi san was interesting for us to ponder what the difference between the two is.

SPECIAL RECOMMEND ▶ 本书特别推荐

中国四大专业盆景网站

请立即登陆

中国岭南盆景雅石艺术网
http://www.lnpjw.com

盆景乐园
http://www.penjingly.com

盆景艺术在线
http://www.cnpenjing.com

台湾盆栽世界
http://www.bonsai-net.com

真柏的取势造型制作

Determination and Modeling of Sargent Savin

改作：樊顺利
文：胡光生
地点：西安中国唐苑
Processor: Fan Shunli
Author: Hu Guangsheng
Place: China Tangyuan in Xi'an city

制作者简介

樊顺利，中国盆景艺术大师、国际盆栽大师。现任中国盆景艺术家协会副会长、安徽省盆景艺术协会常务副会长、世界盆景石文化协会常务秘书长。

About the Creator

Fan Shunli, Chinese Penjing art master; International Penjing master. Now he is the vice president of China Penjing Artist Association; Executive vice-president of Anhui Penjing Artist Association; Executive secretary-general of World Bonsai Stone Culture Association.

图1、2 为改作前的基本树相。该树经初步制作，定势为悬崖造型，现长势旺盛、枝条丰富，为再次创作提供了很大的空间

It shows the basic tree phase before recreation. This tree has been primarily fabricated and the gesture has been determined to be the modeling of cliff. It has grown vigorously by now with abundant branches, which leave a great creative space for the recreation

On-the-Spot 中国现场

图 4 清理完部分枝条后，可以看清树干走势和条位点，这样可依据树干原生枝的内结构，决定下一步造型定势
It is a picture took after some branches are removed. The trunk trend and branch position point can be clearly seen. Thus the next modeling and gesture determination can be decided according to the inner structure of the original branches of the trunk

图 3 在定势前先修剪一些杂乱无章的枝条，使树势和枝、干分布清晰可见，并向唐苑盆景技师们讲解；他在造型前保留和去掉枝条的原因
Cut and trim some rambling branches before gesture deciding, make the tree gesture and the distribution of branches and trunk to be clear and visible, and explain to Penjing artificers of Tangyuan about why he saved or cut the branches before modeling

图 5、图 6 选定角度调整树势，确定主观赏面
Select the angle and adjust the tree gesture, determine the re-modeled main appreciation front

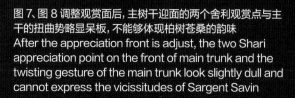

图7、图8 调整观赏面后，主树干迎面的两个舍利观赏点与主干的扭曲势略显呆板，不能够体现柏树苍桑的韵味
After the appreciation front is adjust, the two Shari appreciation point on the front of main trunk and the twisting gesture of the main trunk look slightly dull and cannot express the vicissitudes of Sargent Savin

图9 根据原舍利的木质纹理进行曲线的贯通疏理性雕凿，使曲线舍利干的线条更加流畅，更符合自然舍利的要求，并向唐苑员工们传授制作舍利干的要领
According to the wood texture of the original Shari, conduct the curve penetrating and reorganizing sculpture, improve the smoothness of curved Shari trunk' lines to better meet the requirements of natural Shari, and teach the essentials of Shari trunk fabrication to the staff of Tangyuan

图10、图11 对出盆处弯曲突出部分舍利线条雕凿处理前后的对比效果。雕刻前臃肿呆滞，缺乏变化与力度；雕凿后肌理凸显，线条流畅有劲道
Compare the effects before and after applying the sculpture treatment to the Shari lines at the extruding bend point which stretches off the pot. It was bloated, dull and lacks of variation and power before sculpture, and after sculpture, the muscle is obvious and the lines are smooth and powerful

On-the-Spot 中国现场

图12、图13 对主干内弯处的舍利线条雕凿处理前后对比
Compare the effects before and after applying the sculpture treatment to the Shari lines at the inner bend point of main trunk

图15 开始布局绑扎枝条,与唐苑盆景师傅们商讨该树的枝条定位点与枝条处理的基本手法、明确了主题指导思想
Begin to set the layout and bind the branches, discuss with Penjing artificers of Tangyuan about the branch positioning point and the basic method for branch treatment, and define the theme guiding thought

图14 在枝条布局造型前,为充分体现主干的曲线变化和透视感,对主干进行适度调整
Before the arrangement and modeling of branches, to fully express the curve variation and perspectivity of the main trunk, properly adjust the main trunk

图16 唐苑盆景师傅们动手用铝丝绑扎枝条
The Penjing artificers of Tangyuan wind the Aluminium wires to the branches

图17 调整好前飘枝
Adjust the front flying branch

图18 翻盆处理
Apply the treatment of pot replacement

图19、图20 翻盆角度调整后，主干前后曲线脉络明显通畅，扭曲、灵动而不失自然效果
After the adjustment of pot replacement angle, the front and back curve context of main trunk are obviously smooth, it twists flexibly and naturally

On-the-Spot 中国现场

图21、图22 改作完成后的背面和侧面
It shows the back side and profile after recreation

原桩坯

图23 为改作后的正面效果图。制作改变了原树势取半悬崖造型之式，其险势依旧，但整体更为轻灵飘逸，主干的九曲回旋得以突显其树材美更为充分的展露，树势比例也很得体
It is the front effect picture after the recreation. The fabrication changes the style of the original tree gesture of semi-cliff modeling. The perilous posture remains the same but the whole gesture seems more ethereal and elegant. The nine twists of main trunk has highlighted and fully revealed the beauty of tree material, and the scale of tree gesture is very appropriate

更正

《中国盆景赏石·2013（古镇）中国盆景国家大展全景报道特别专辑》中《展示松树造型艺术的正能量——樊顺利大师在2013（古镇）中国盆景国家大展上的现场制作表演》一文的图10、图11（P86）的图注有误。图11的手绘图并非樊大师所画，而是日本森前诚二先生在看到原素材时个人所绘草图，在樊大师完成改作后，森前先生拿出图纸进行对比，发现二人的思路竟不谋而合。特此说明并更正。

Heart for Creation of Artworks
- Recreation of "Breeze", a Famous Juniperus Chinensis Penjing

创作艺术作品之心
——真柏盆景铭品"清风"的改作

撰文、制作：【日本】小林国雄　图注：茗源山人
Author & Artist: [Japan] Kunio Kobayashi　Pictures: Shaoyuan Recluse

作者简介

小林国雄，日本水石协会理事长、春花园 BONSAI 美术馆馆长、《中国盆景赏石》海外名誉编委兼顾问、日本盆栽协会公认讲师。

曾 4 次获得日本盆栽作风展内阁总理大臣奖，6 次获得皋树展大奖。有 16 人通过他亲手创作或指导下的盆栽获得国风奖，并有 200 人次以上入围国风盆景展。获得由日本文化振兴会颁发的"国际艺术文化奖"、中国盆景艺术家协会颁发的"中日盆景文化交流大使"、日本江户川区颁发的"文化功绩奖"等多个奖项或称号。

作为日本代表性的盆栽作家，他曾在东京都会电视台《盆栽》节目担任讲师长达 2 年半，并积极到世界各国进行盆栽的示范表演、演讲和做评委等，所到国家多达 20 个以上，受到广泛欢迎。著作《盆栽》被译为 7 种语言，成为远销海外的畅销书；随笔《口八丁手八丁（能说善做）》在杂志中连载长达 3 年；另有著作《盆栽艺术——天》、《盆栽艺术——地》等。

Author Introduction

Kunio Kobayashi, the director general of Nippon Suiseki Association, the curator of Syunkaen BONSAI Museum, the international honorary editor and consultant of "China Penjing &Scholar's Rocks", the well-known lecturer of Japanese Bonsai Association.

He is the 4-times winner of Prime Minister Prize at Japan Bonsai Sakufu-ten Exhibition; 6 times-winner of Grand Prix at Kojuten Azalea Exhibition. Under his schooling, 16 students and enthusiasts have been awarded Kokufu-sho Prize and 200 have been awarded a prize at Kokufu-ten Bonsai Exhibition. He was awarded International Art &Culture Prize by Japan Culture Foundation; Sino-Japan Penjing Culture Ambassador by China Penjing Artists Association; honored with Contribution to Culture Award by the mayor of Edogawa Ward and so on.

As the representative bonsai creator, he hosted a weekly show "BONSAI" on MX YV for 2 and a half years. He traveled to more than 20 countries to give lectures and demonstration; also he often takes the role as judges. He is widely welcomed.

His writing BONSAI has been translated into 7 languages and become a overseas best-seller; he contributed an essay titled "Kuchi Haccho, Te Haccho-As Ready in Talk As in Deed" to a magazine for 3 years; published "BONSAI GEIJYUTSU-Ten-", "BONSAI GEIJYUTSU-Chi-" and so on.

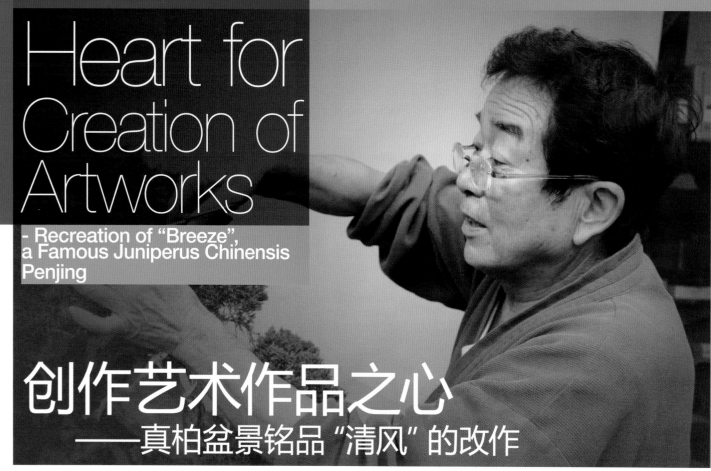

图1 真柏盆景"清风"，是小林国雄先生的盆景佳作之一，这件作品曾经参加过第十七回"日本盆栽大观展"，并荣获"内阁总理大臣赏"，这是当年获奖时的样貌

The juniperus chinensis Penjing "Breeze" is one of the best Penjing works of Mr. Kunio Kobayashi. It has been exhibited in the "17th Nippon Bonsai Taikan-ten Exhibition", and won the Prime Minister Prize. The picture was taken when the prize was awarded

Spot International 海外现场

图2 随着时代的发展,审美意识也在不断发展变化。对于这件作品现在的模样,小林国雄先生已感不满,决定进行一次改作,以期提升该作品的品格
With the development of time, the aesthetic consciousness also develops unceasingly. Mr. Kunio Kobayashi was not satisfied by the current appearance of the artwork and decided to recreate it and enhance its quality and concept

图3 原作的枝叶过于繁密,首先是删减过多的枝叶
The branches and leaves of the original work are too dense, the first step is to eliminate the overmuch branches and leaves

图4 学生将剪除树叶后留下的小枝剥去树皮,做成舍利枝,作虚化处理
The students cut of the leaves from the branches and peel the bark off, made it as Sheri branch and performed the blurring treatment

我作为盆栽作家已经近40年光景,这期间我始终在探索"盆栽之美的本质是什么"。园艺与盆栽同样是把植物种在盆钵之中。园艺是通过欣赏花、叶、果实的美,以其华美缤纷的外在美来治愈心灵。但与之相反,盆栽是感受到内心中深藏着的物哀的生与死,经过时间的流逝感受到幽静、闲寂之后,心灵会被时代酝酿出的意境韵味所吸引,会被生命的尊严所感动。

文化勋章的受奖人——雕刻家朝仓文夫如是说,"艺术创作是由才能的训练而产生的。艺术家并不是知识分子,即便给予一个人知识他也成不了艺术家,对于艺术家来说,才能升华的基础在于坚持不懈地做自己喜爱的事情。"

盆栽和雕刻都是创作出立体的造型,在这点上是非常相似的。但是,雕刻或者盆栽只是模写出精巧的外形是不能成为艺术品的。必须要表达出对象本身具有的独特的精神、风格之美。雕刻被称作是,为挖掘出隐藏的内在美、捕捉到其灵魂,而去除多余部分的工作。盆栽亦与此相同,去除素材原本多余的部分,耐心培育植物的生命力,进行整形,使自然美栩栩如生地展现出来。

图5 对作品的顶冠部分作细心地删剪
Carefully eliminating and trimming the crown part of work

图6 对学生们讲解删剪的方法,用原作的图片对比剪后的效果
Explain the eliminating and trimming methods to the students and use the picture of original work to compare with the effect after trimming

图7 仔细地审视作品，思考树枝进一步的删剪去留，将考虑剪除的树枝先用布匹遮盖起来，以便更直观地看到效果，做到万无一失后才下剪
Carefully review the artwork, think about how to further eliminate and trim the branches. Cover the branches planned to be cut off with clothe for the convenience to see the effect directly, and then cut the branches off after full preparation

图8 删剪完成后，将留下的枝叶用金属丝攀扎调整
After the cutting and trimming are done, use the metal wire to bind and adjust the left branches and leaves

图9 将雕刻后的舍利枝涂上石硫合剂，作防腐处理
Paint the sculptured Sheri branch with lime sulphur for preservative treatment

图10 舍利干的基部近盆面处，由于长期处于潮湿状态，其中部已渐渐腐朽而中空，具有古柏透瘦苍古的美感。但如果腐朽殆尽，则会"过犹不及"，影响树的稳定感和审美价值，所以必须经常涂以石硫合剂作防腐处理
For the base part of Sheri stump near to the pot surface, due to the humid situation it has been in for long time, the central section has gradually become corrosive and hollow, and that reveals the emaciated and antique beauty of ancient juniperus. However if it is wholly corrosive, the result will be "Overdone is worse than undone." and affects the stability and aesthetic value of the tree. Therefore, the lime sulphur should be often applied for preservative treatment

为了把具有生命的盆栽提升到艺术作品的高度，对超越目标的人之本性进行深度陶冶是必要的。

这次，我在2013年第22届日本盆栽作家协会展中参展的真柏，是纪州的下山桩，树龄有500年。在1998年的第17届日本盆栽大观展中参展并获得内阁总理大臣奖。那时的我局限于仅仅基本上执著地、漂亮地完成创作而已。在参观中国的盆景时我感受到了重要的东西。近年来日本的盆栽作品的最终作品风格都是被漂亮地整理成划一的造型。但是中国盆景的作品风格却是让树充分发挥出本身的个性。从中国岭南派盆景强有力的线条的动势中，我学习到了盆栽之美的本质是什么。

这次我对真柏的改作，是从园艺到盆栽、从面到线的改作。大胆地剪掉漂亮而细致整齐的簇状枝，彰显出线条的跃动感和空间感，展现出大自然的残酷意境和韵味。

展览后，为追求更美的境界，我在舍利干上开洞以展示出残酷性。作家如果没有对美进行深入地研究，就无法创作出感动心灵的作品。因此，以日以继夜、不辞辛劳的精神去钻研是十分必要的。

最后，在中国古典著作《周礼》的《考工记》中记载着创作的四个条件，即"天地材工"。

天有时，地有气，材有美，工有巧。合此四者，然后可以为良。

（编辑注：本文中的园艺应该是尤指盆花、盆草、盆树等的栽培）

图11 树冠的删剪整理已经完成，将作品擦拭的干干净净，这是对作品及自我的一种尊重，也是作者创作精神的体现
The cutting and trimming procedures for the tree crown have been completed, wiping the artwork clean is a respect for the work and artist himself, it is also the expression of the artist's creation spirit

Spot International 海外现场

I have been a bonsai artist for nearly 40 years, and during this period, I have been searching for the answer to "what is the nature of bonsai beauty". Both horticulture and bonsai are to cultivate plants in pots. The horticulture is to cure the heart with the gorgeous and colorful outer beauty by appreciating the prettiness of flowers, leaves and fruits. However as the contrary, the bonsai is for people to feel the deeply hidden life and death of sentimentality. As time goes by, and after the feeling of silence and vanity, the heart will be attracted by the artistic conception and lasting appeal brewed by the age and touched by the dignity of life.

The awardee of Medal of Culture– the sculptor Fumio Asakura has said that "The artistic creation is generated by skill training. The artist is not an intellectual. One cannot be an artist if he is educated only with knowledge. For an artist, the talent sublimation is based on unremittingly doing what he is enthusiastic about."

Both the bonsai and the sculpture are very similar in creating a stereo modeling. However, the sculpture or bonsai are just copies of the elaborate appearance, which cannot become an artwork. It has to express the unique spirit and beautiful style of the object itself. The sculpture is known as the work of excavating the hidden inner beauty, catching the soul and eliminating the overmuch part. The bonsai is also the same. The artist eliminates the overmuch part of the material, patiently cultivates the vitality of plants, performs the modeling and reveals the lifelike natural beauty.

图 12 删剪重整后的真柏盆景,比删剪前更具古柏的样貌
The trimmed and remodeled juniperus chinensis Penjing is more like the ancient juniperus than before

图 13 纤细的舍利枝与苍翠的绿叶相映,平添了虚实变化之美
The slender Sheri branch and the verdant green leaves highlight each other and add the beauty of change between vacancy and reality to the artwork

In order to enhance the bonsai with life to the height of artwork, it is necessary to deeply cultivate the human nature of exceeding goals.

At this time, my juniperus chinensis bonsai exhibited in the 22th Japan Bonsai Sokufuten Exhibition is the lower mountain-like stump in Kishu and with the tree age of 500 years. It has been exhibited in the 17th Nippon Bonsai Taikan-ten Exhibition in 1998 and won the Prime Minister Prize. And at that time I was limited only to basically finish the creation tenaciously and nicely. I have felt something important when appreciating China Penjing. In recent years, the final artwork style of Japanese bonsai is prettily trimmed to the same modeling. But the artwork style of China Penjing is to fully express the own personality of tree. From the momentum of strong and powerful lines of Chinese Lingnan Penjing, I've learnt what is the nature of bonsai beauty.

图 14 删剪后,树枝的线条也得以展露出来,呈现扭转变化的效果
After eliminating and trimming, the lines of branches are revealed and appears the effect of twisting change

图15 与学生们在作品前合影,共享成功的喜悦
Taking photo with the students before the artwork and sharing the happiness of success

My recreation of juniperus chinensis at this time is a recreation from horticulture to bonsai, and from plane to line. I daringly cut off the prettily and finely organized clustered branch to highlight the senses of leaping and space and express the artistic concept and the appeal of cruelty of nature.

After the exhibition, in order to pursue a more beautiful state, I picked holes in Sheri stump to show the ruthlessness. If the artist does not make deep research on beauty, he cannot create the artwork of heart inspiration. Hence, it is very necessary to study intensively with tireless spirit night and day.

At last, 4 creation conditions recorded in Kaogongji of Zhouli, the Chinese classic work, are "weather, earth vitality, material and skill".

The four elements of appropriate weather, vital earth, beautiful material and elaborate skill can form a great piece of work together.

(Notes of the editor: The horticulture indicated in this article shall especially mean the cultivation of potted flowers, potted grass, and potted trees, etc.)

图 16 再度审视经改作的真柏盆景,似觉还有地方不尽如人意。原来是主干中部的舍利过于宽阔平板,与基部透瘦的舍利不协调,还得再次进行雕凿加工
Review the recreated juniperus chinensis Penjing again, and seems that some parts are still dissatisfactory. It turns out that the Shari at the central part of main stump is too broad and flat and inharmonious with the skinny Sheri at the base part. The sculpture processing shall be conducted again

图 17 将主干舍利最宽阔处的右侧作洞穿通透处理
Perform perforating and transparency treatments at the right side of Sheri of the main stump

图 18 对雕凿处进行打磨
Polishing the sculptured area

Spot International 海外现场

芸術作品を創る心
——真柏銘樹「清风」の改作

実技・文：〔日本〕小林國雄　写真の文：茗源山人

图19 经雕凿打磨后的效果
The result after sculpturing and polishing

图20 将原本较平面的舍利刻出立体而变化的线条
Carve the Sheri which is originally flat in stereo and changeable lines

作者紹介

小林國雄氏は、(社)日本水石協会理事長、『春花園BONSAI美術館』館長、『中国盆景賞石』海外名誉編者兼顧問、日本盆栽協会公認講師。日本盆栽作風展の内閣総理大臣賞を4度受賞、皐月展大賞の回受賞。また、国風盆栽展「国風賞」受賞者16人、入選者200人以上の指導・盆栽を手がける。日本文化振興会より国際芸術文化奨励賞受賞、中国盆景芸術家協会より中日盆景文化交流大使受賞、江戸川区より文化功績賞受賞。東京MXテレビ『BONSAI』に講師として2年半レギュラー出演。また、日本を代表する作家として、海外の20ヵ国以上からも実技演習、講演、品評で引っ張りだこの人気アーティスト。世界7ヵ国で翻訳されベストセラーになっている『BONSAI』著書には、エッセイ『口八丁手八丁』3年連載。作品集『盆栽芸術 天』『盆栽芸術 地』などもある。

私は盆栽作家として四十年近く「盆栽の本質の美とは何か」を模索し続けて来た。園芸も盆栽も植物を鉢植えにすることに於いては同じ。園芸は花や葉や実の美しさを鑑賞し、華美・華飾的な美しさによって心が癒される、反面盆栽は内奥に秘められたものあわれの生死を感受する。長い時間経過による侘び寂び、時代が醸し出した風趣風韻に心が魅せられ、生命の尊厳に感動させられる。

文化勲章を受章した彫刻家の朝倉文夫氏は言う。「芸術創作は才能の訓練から生まれる、芸術家は知識人ではない、知識を与えても芸術家はできない、芸術家に必要な才能の発達は自分の好きなことをやることによって促される」。

盆栽と彫刻とは形を立体的に造り出すことに於いて、もっともよく似ている。しかし彫刻も盆栽もただ形を精巧に写すだけでは芸術作品とは言えないのである。対象そのものが持つ個々それぞれの精神・風格が美しく表現されなければならない。彫刻は対象の内部に秘められている美を掘り起こすために、その魂を捉えて余計なものを取り除き仕事だと言う。盆栽もそれと同じで、素材の持つ余計なものを取り除き、植物の成長力を気長に育み、形を整え、自然美を生動させることに於いて一致している。

生命ある盆栽を芸術作品にまで高めるためには、対象に挑む人間性の陶冶高揚を極めることが必要とされる。

今回、私が二○一三年の第二十一回日本盆栽作家協会展に出品した真柏は、紀州産の山取りで、樹齢は五百年はたっている。一九九八年の第一七回日本盆栽大観展に出品し内閣総理大臣賞を受賞している。その頃の私は基本に固執し、綺麗に仕上げることにだけ囚われていた。しかし、何度も中国に訪れ、中国の盆栽を観ている時に重要なことに気付かされた。近年の日本の盆栽作品は綺麗に整えられた画一的な姿に仕上げられた作風だが、中国の盆栽作品は樹そのものが持つ個性を生かした作風である。中国の岭南盆景の力強い線の動きから、私は盆栽の本質の美とは何かを学んだのである。

今回の真柏の改作を見ていただきたい。園芸から盆栽に、面から線への改作である。綺麗に几帳面に整えられた枝棚を大胆に切り落とし、線の動きと空間を引き出した。大自然の厳しい風趣・風韻を引き出したのである。展示後、さらになる美を求め舎利幹に穴を開けて厳しさを現出した。作家は美に対する「業の深さ」がなければ人の心を感動させる作品は創れない。そのためには日々のたゆまぬ精神の研鑽が必要なのである。

最後に、中国の古典「周禮」の経典の中の考工記にモノ創りについての四条件「天地材工」がある。

天に時有り 地に気有り
材に美有り 工に巧有り
此の四者を合わせ 然る後
可以て良と為す

图21 进一步的加工制作完成后,舍利更苍古而自然,与散落纷披、虚实相间的树枝相协调
After the further processing and fabrication are completed, the Sheri seems more ancient and natural, and coordinating with the scattered branches showing mixed vacancy and reality

图22、23 早在2500年前的春秋时期,中国的先哲著有《考工记》一书,认为天时、地气、材美、工巧是设计制作优秀物品的四个要素,这一观点至今仍为重要,这件真柏盆景的改作,正是契合了这一理念
Early in Chunqiu Period before 2500 years, a Chinese sage has written the work of Kaogongji, he believed that the appropriate weather, vital earth, fine material and elaborate skill are the four elements to design and fabricate an excellent object. And this opinion is still very important in nowadays. This recreation of juniperus chinensis Penjing has just corresponded to this concept

Spot International 海外现场

图24 让我们再来对比看一看作品改作前后的模样
Let's see the appearances of artwork before and after the recreation in comparison

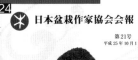

图25 中国有句古语叫做"谨毛失貌"，意谓艺术创作时刻意追求细部效果，就会失去整体的大效果、大气象。真柏盆景铭品"清风"的改作，正是舍去了原本枝叶精细严整的"微毛"，却获得了神韵气象的"大貌"，将作品向时空深处推进了千百年，因此取得了改作的成功
There is an old saying in China as "Being cautious about small part might cause overall loss," which means that deliberately pursuit for the detailed effect during artistic creation will cause loss in the overall effect and the spirit of work. The recreation of the famous juniperus chinensis Penjing "Breeze" has eliminated the "small parts" of branches and leaves which are originally elaborate and strictly tight, and gained the "overall magnificence" of the romantic charm and scene. It has pushed the artwork for thousands of years in the sense of time and space, and that's how the recreation gets succeed

唐苑的世界盆景对话

SPECIAL ISSUE OF "2013 DIALOGUE TO THE WORLD PENJING"

特别专辑

本栏目通过发言人提供的文字、图片、论坛当天录音整理而成

2013 唐苑的世界盆景对话
DIALOGUE TO THE WORLD PENJING

寄情树石 求索真谛
Love in Trees and Stones Exploration for Truth

文：王选民　图文整理：CP　Author: Wang Xuanmin

今天众人探索盆景赏石之道，寻找自然美之根源和古人精神世界之奥秘。让盆景人得到其中的乐趣，确实是一件幸事，也是一件善事。

纵观中国几千年的文化历史，可以肯定地说人与自然景物的关系最为密切。无论是物质需求还是精神需求，都要建立在人与物的关系上。就盆景而言，人们是借助于一棵树一块石头去认识它、发现它的本体特征和天性，从中产生美的评价，然后将自己的思想情感寄托于树石之中。在作家创作及表达过程中，最终将主体思想对象化。此时，树的形象特征已经人格化了。艺术实践证明，就创作目的而言作家所追求的不单纯是一个形象而已，最重要的是在借助于物体形象的建立，从中去表达探索寻找某一种心理上的思想境界。例如：树形所产生的气势、神姿、神彩、神韵、气息、气韵生动等。

这种美意识的产生是在有限的有形的存在中，去探索感受无限的无形的存在，有时它能超出我们的正常理解，摸不到但可以意会到！意会到了可以让人"动心"，甚至触及灵魂！这就是我们所说的意境、幽玄之境或者是与物神交的效果吧！

编按：王选民老师因故未能参加论坛活动，亦是发来信息，与世界盆景人共同探讨。

王选民，中国盆景艺术大师，BCI国际盆栽俱乐部理事。现任中国盆景艺术家协会名誉副会长，世界盆景石文化协会理事。
Wang Xuanmin, China Penjing Art Master, director of BCI (International Bonsai Club). Now, Wang Xuanmin is the Vice-president of Chinese Penjing Artists Association, director of World Bonsai Stone Culture Association.

Today, we all are exploring the way how to appreciate Penjing to find out the root of natural beauty and the secret of the ancients' spiritual world. To let Penjing people find pleasure in Penjing appreciation which is indeed a good and happy thing.

Throughout Chinese thousands of years of cultural history, we can say for sure that the relation between man and nature is most close. Whether material requirements or spiritual requirements, they are based on the relationship between humans and things. For Penjing, people know it and find its natural characteristics and instincts by means of a tree or a stone, and then have the judge of beauty and place their thoughts and emotions on the tree or stone. In a Penjing maker's creation and expressions, finally the Juche idea will be objectified. Right now, the image features of the tree is being personalized. Artistic practice has proved, as far as the creation purpose is concerned, that what the maker is seeking for is not simply an image, but more importantly to explore an ideological level psychological by means of the establishing of an object image, such as the temperament, gesture, color, charm, vitality and vigor from the shape of the tree.

Such generation of aesthetic sense is explored and felt in limited and physical existence. Sometimes, it will go beyond our normal understanding, as a result that we can't touch it but can feel it. Feeling it will make us "moved" and reach its soul. This is what we call the artistic conception, mysterious state or the effect of spiritual mixing with things.

Editor's Note: Prof. Wang Xuanmin is unable to participate in the activities of the Forum, but send messages to discuss with the World Penjing people.

中日盆景交流
——对须藤雨伯所提问题的回答

2013 唐苑的世界盆景对话
DIALOGUE TO THE WORLD PENJING
—Reply the Questions from Mr. Sudo Uhaku

发言人：徐昊 图文整理：CP Speaker: Xu Hao Reorganizer: CP

盆景是起源于中国的一项艺术，在古老中国的土地上盛行已久，今天在美丽的唐苑举办的"唐苑的世界盆景对话"国际年度论坛意义非凡。

在这里，我要感谢日本的盆景人，近一百多年来，是他们将盆景艺术传遍全球，成为一项世界性的艺术。今天，在此就须藤雨伯先生提出的一些问题和大家进行交流。

一、在中国，盆景被继承下来已有2000多年的历史，它能被传承如此之久的原因是什么？
1. 源自心灵深处的热爱。
2. 农耕文明为盆景艺术的发生和发展提供了条件。
3. 盆景艺术根植于中国传统文化。

二、中国盆景为什么存在？（存在的目的是什么？）

有需要就会有存在。在唐代，诗人白居易就以诗歌的形式将盆景的好处归纳为10条。诗中写道："养性延容颜，助眠除睡眠。澄心无秽恶，草木知春秋。不远有眺望，不行入洞窟。不寻见海埔，迎夏有纳凉。延年无朽损，升之无恶业。"这是诗人心目中当时盆景存在的意义。

盆景艺术的存在，主要取决于人们的精神需求和物质需求两个方面，但最主要的是精神层面上的追求。其主要目的表现在如下三个方面：
1. 借物言志。
2. 审美需求。
3. 物质意义。

三、在中国2000年的历史中，盆景充当了怎样的角色（指唐朝～清王朝）？
1. 审美对象。
2. 寄情之物。
3. 文化的载体。
4. 养生之道。

四、关于中国盆景的山水思想，特别是道教中的"气"和"不老长寿"？
1. "日与自然相亲和"——中国盆景的山水思想。
2. "虚处即妙境"——道家文化中的"气"在盆景中的反映。
3. 长寿即是"神仙"——"不老长寿"思想在盆景中的反映。

五、中国盆景所说的"意境"是什么（意境的重要性）？

"诗情画意"即意境——中国盆景的意境。

意境是超越具体有限的物象，让心灵深入时空的运动和感悟，是心里所想而现实却达不到的场景。

意境是盆景作品的灵魂，没有意境的作品便是一件没有灵魂的躯壳，只仅仅是自然植物给视觉带来的愉悦。

六、"外师造化，中得心源"——关于艺术创作和自然造化之美？

中华文化艺术历来崇尚自然，主张"外师造化"，外师造化就是向大自然学习。中国盆景艺术不仅仅是"外师造化"的客观表现，更注重"中得心源"的主观表达（"心源"即心底深处的精神情感世界）。

七、对于自然的摹写是什么（树姿、树相、树灵）？

艺术创作的根源，是来自于心灵对自然和生活的感动，从而以某种形式为载体，把内心的感动表达出来。

编按：《中日盆景交流——对须藤雨伯所提问题的回答》（简称《回答》）是中国盆景大师徐昊先生对日本盆栽大师须藤雨伯先生在《中国盆景对日本盆栽的影响（作品风格·审美意识）》中所提问题的回应，两篇文章均已成稿，以中英日三语刊登在《中国盆景赏石·2013-10》P54～P65页上，因版面所限，在此只摘取《回答》一文的提纲。

The Annual Forum 国际年度论坛

徐昊，中国盆景艺术大师，BCI国际盆栽大师。现任中国盆景艺术家协会第五届理事会副秘书长兼会长助理。
Xu Hao. China Penjing Art Master、International Bonsai Master by BCI. Now, Xu Hao is the Deputy secretary general and President assistant of Chinese Penjing Artists Association 5th Executive Committee.

Penjing is an art originated from China, it has been popular in the ancient land of China. It is so meaningful to hold the "the dialogue to the world Penjing" at the beautiful Tangyuan.

Here, I want to thank Japanese Penjing people who has been promoting Penjing art to the whole world and making it become a worldwide art. Today, I will discuss with all of you about Mr. Uhaku Sudo'question

I. In China, Penjing has been carried on over 2,000 years. Why the inheritance of Penjing can be kept for such a long time?
1. Because of the love for Penjing living deep in the heart.
2. Agriculture civilization provides conditions for generation and development of Penjing art.
3. Penjing art is rooted in Chinese traditional culture.

II. Why does China Penjing exist (what is the purpose of its existence?)

Where there is need there is existence. In Tang Dynasty, poet Bai Juyi summarized ten advantages of Penjing using the form of poem. The poem writes, "Cultivate nature & maintain youth, eliminate fatigue & assist sleep. Purify soul without evilness, judge the season by the growth of grass and trees. See the whole scene without standing in a far place and just like place oneself in a cave without walking. Without seeing Haipu, enjoy the cool in summer. Remain intact for a long time, having no evil deed when raise the bar." This is the significant existence of Penjing in the poet's eye at that time.

The existence of Penjing art mainly depends on human spiritual and material needs with the former as the leading part. The main purpose is as follows:
1. Making use of objects to express ideas.
2. Aesthetic needs.
3. Material significance.

III. What role has Penjing played through China's 2,000-year history (refer to Tang Dynasty—Qing Dynasty)?
1. Aesthetic object.
2. Pinning feelings on objects.
3. Carrier of cultural .
4. The art of living.

IV. Landscape thought for China Penjing, especially "Qi" and "immortality" of Taoism
1. "Love nature"—landscape thought of China Penjing.
2. "Unreality is a wonderland"—"Qi" in Taoist culture reflected in Penjing.
3. Longevity is "immortal"—reflection of "immortality" thought in Penjing.

V. What does "artistic conception" of China Penjing refer to (the importance of artistic conception)?

"Idyllic" is artistic conception—artistic conception of China Penjing

Artistic conception is the image exceeding specific and limit, the motion and inspiration enabling the soul go deep into time and space and a scene strived for but could not be reached in reality.

Artistic conception is the soul of a Penjing work. A work without artistic conception is nothing merely an outer form without soul, which is just the visual joyfulness brought by natural plants.

VI. "Art creation originates from masters' thoughts for nature with essential inner inspiration"----the beauty of art creation and nature?

Chinese culture and art always advocate that nature provide people with inspiration which refers to learning from nature.

China Penjing art is not only the objective representative of "masters' thoughts for nature", it also emphasizes subjective expression of "getting inner inspiration" ("Inner inspiration" is the spiritual and emotional world in deep heart).

VII. What are depicted for nature (tree performance, tree appearance and tree soul)?

The source of creation originates from soul touched by nature and life, therefore art creation expresses inner feelings in different forms.

Editor's note: "China-Japan Penjing Communication — Reply the Questions from Mr. Sudo Uhaku"(hereafter referred to as "Reply") is the reply of China Penjing master Xu Hao to Japanese Bonsai master Uhaku Sudo'question in the article "The Influence of China Penjing on Japanese Bonsai (Characteristic Style of Works, Aesthetic Consciousness)". The two article has been published on "China Penjing&Scholar's Rocks" 2013-10 from page 54 to page 65 in three languages-Chinese, English and Japanese. Due to the limited pages, here we only extract the syllabus of the "Reply".

2013 唐苑的世界盆景对话
DIALOGUE TO THE WORLD PENJING

美国的盆景

发言人：【美国】威廉·尼古拉斯·瓦拉瓦尼斯 图文整理：CP

威廉·尼古拉斯·瓦拉瓦尼斯《国际盆栽》杂志出版人兼主编
William Nicholas Valavanis Publisher and Editor of International BONSAI Magazine

日本细齿冬青树
培育：布恩·曼妮凯蒂帕特
Japanese Fine-Tooth Holly.
Trained by Boon Manikitivipart

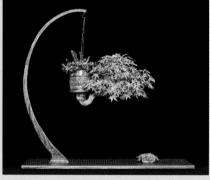

鸡爪枫
培育：约瑟夫·诺加
Japanese Maple.
Trained by Joseph Noga

早期的拍卖和展会

盆景最初是作为礼物或者以公开拍卖的形式被引入美国的。1863年，亚伯拉罕·林肯总统派遣蒲安臣作为驻华大使。在经过日本时，当时的天皇将一盆橡树盆栽赠予大使作为礼物。大使将盆景送到加利福尼亚州，这盆盆景现在收藏于坐落在加利福尼亚州奥克兰梅里特湖畔的金国盆景联合会北收藏馆。或许这是至今为止美国最早的有记录的盆景。在美国，最早的盆景公开展览在1876年的费城百年纪念博览会上。1893年在芝加哥伊利诺斯州举办的哥伦布世界博览会上也有盆景的展示。在东北部地区，盆景也出现在一些拍卖会上。1904年的5月4日～5日，纽约市举办了一场名为"非凡的日本植物"的拍卖会。

盆景的先驱们

约翰·纳卡是一名在美国出生的盆景艺术家，早年他与其祖父一起在日本生活并在那儿结识了盆景。回到家后，他从事园艺工作并开始以美国的本土树种来进行盆景的种植。纳卡先生周游美国和世界各地传授他的技巧，但更重要的是在每一个国家推广使用本土植物作为盆景的素材。他有两本主要的盆景著作：《盆栽技术 I》和《盆栽技术 II》。

美国国家盆景和盆景博物馆坐落于华盛顿的美国国家植物园内。由于约翰·纳卡先生对盆景做出的杰出的推广工作，北方美国盆景收藏馆的建筑以约翰·纳卡的名字命名。我相信这是第一次以一位在世的人的名字对国家建筑物进行命名。

吉村裕治是一名来自日本东京的第二代古典盆栽艺术家和学者，他从1952年便开始向公众传授盆栽知识，也是第一位这样做的人。1958年，他移民到美国传授盆栽知识并成立了吉村盆栽公司，后改名为吉村盆栽学校。他参与了包括《微型树木和风景之日本艺术》等书籍的创作，而这本书也是第一本英文版的关于盆栽的综合性著作。

除了在全美范围内教授传统盆栽知识外，他还游历了澳大利亚、英国、印度和中国香港向人们介绍他的艺术。而他的艺术则传承于他的父亲——吉村俊治，日本一位卓越的盆栽和水石权威。在华盛顿的国家盆景和盆景博物馆内，以他的名字命名了吉村裕治教育中心。

盆景收藏馆

在美国有几个主要的盆景收藏馆。最古老的是位于马萨诸塞州波士顿的阿诺德植物园的拉兹·安德森盆栽收藏馆。拉兹·安德森曾经是一名驻日大使,并在1913年引进了盆栽。他的遗孀在1937年向阿诺德植物园捐献了30盆盆栽,由此便成立了拉兹·安德森盆栽收藏馆。

位于纽约布鲁克林的布鲁克林植物园拥有早期的盆景收藏并且出版了一些深受大家喜爱的阅读手册。植物园还向公众开放了盆景讲座班。1957年,吉村裕治初到美国时便在这里授课并发展了此处的盆栽收藏。

坐落于华盛顿美国国家植物园的美国国家盆栽和盆景博物馆作为如今最好的博物馆致力于在世界范围内推广盆景。全年都举办培训班、季度博览会以及其他形式的活动。该博物馆藏有日式盆栽收藏、中国盆景收藏、北美盆栽收藏以及热带盆栽收藏。

环太平洋盆栽收藏馆位于西雅图附近费德勒尔韦的惠好校区。该收藏馆藏有来自于太平洋附近国家的盆景。

位于加利福尼亚州的金国盆栽联合会拥有两个著名的盆景收藏馆。北部收藏馆坐落于加州奥克兰的梅里特湖畔,而南部收藏馆则位于加州帕萨迪纳市的亨廷顿花园。

盆景活动

在美国有着大型的盆景社团,在广阔的国土上分布着200多个盆景俱乐部。在加利福尼亚、德克萨斯、佛罗里达、新泽西和纽约都举办有年度盆景大会。

虽然美国盆栽协会的成员分布世界各地,但是大多数还是来自美国、加拿大和墨西哥。他们出版了名为《盆栽》和《美国盆栽》协会杂志的季刊,其中的技术性文章引起人们对盆景栽培的兴趣,帮助人们创作更好的盆景。

《国际盆栽》是第一个也是唯一的一个在美国出版的专业的盆景杂志,并且包括了来自日本以及世界各地的学术性文章。本杂志已经出版了36个年头,杂志为所有盆景期刊订立了指向标。

大多数盆景协会每年至少举办一次盆景展会以向公众推广盆景并分享他们的盆景作品。

在美国,人们对盆景的兴趣是积极踊跃并且是不断增长的。

三角枫 *Acer buergerianum*
培育:苏鑫·苏克洛梭维斯特
Trident Maple. Trained by Suthin Sukolosovisit

桧柏,即圆柏 *Sabina chinensis*
培育:约翰·纳卡
China Savin. Trained by John Y. Naka

圆柏 *Sabina chinensis*
培育:布恩·曼妮凯蒂帕特
China Savin. Trained by Boon Manikitivipart

圆柏 *Sabina chinensis*
培育:布恩·曼妮凯蒂帕特
China Savin. Trained by Boon Manikitivpart

The Origin & History
of Penjing in the United States

Speaker: [America] William Nicholas Valavanis Reorganizer: CP

Early Auctions and Exhibitions

Penjing were first introduced into the United States as gifts or at public auctions. President Abraham Lincoln appointed Anson Burlingame as his envoy to China in 1863. While passing through Japan the Emperor presented him with a Daimyo oak bonsai. He sent the Penjing to California and it's now in the Golden State Bonsai Federation Collection North at Lake Merritt in Oakland, California. Perhaps this is the oldest documented Penjing in the United States today. The first public display of Penjing in the United States was at the Philadelphia Centennial Exposition held in 1876. Penjing were also shown at the 1893 Columbian World Fair held in Chicago, Illinois. Several auctions included Penjing in the Northeast. An auction of Remarkable Japanese Plants was held in New York City on May 4-5, 1904.

Penjing Pioneers

John Y. Naka was an American born Penjing artist who lived with his grandfather and got acquainted with bonsai in Japan during his early years. Returning home he worked as a gardener and began growing Penjing with native Amerian species. Mr. Naka traveled around the United States and the world teaching his techniques, but more importantly achievement is to promote the use of native plant material for Penjing in each country. He authored two major Penjing books, *Bonsai Techniques I* and *Bonsai Techniques II*.

The National Bonsai and Penjing Museum is located on the grounds of the U.S. National Arboretum in Washington, D.C. John Naka was honored for his work promoting Penjing by naming the complex where the North American Bonsai Collection is housed as the John Y. Naka North American Bonsai Pavilion. I believe this is the first time a national building was named for a person who was alive.

Yuji Yoshimura was a second generation classical bonsai artist and scholar from Tokyo, Japan, who was the first to teach bonsai to the public, beginning in 1952. He immigrated to the United States in 1958 to teach bonsai and began the Yoshimura Bonsai Company which was later renamed the Yoshimura School of Bonsai. He co-authored several books including *The Japanese Art of Miniature Trees and Landscapes*, which is the first comprehensive bonsai book in English.

In addition to teaching classical bonsai throughout the United States, he traveled to Australia, England, India and Hong Kong to introduce his art which was learned from his father, Toshiji Yoshimura, a prominent bonsai and suiseki authority in Japan. The Yuji Yoshimura Educational Center was named in his honor at the National Bonsai and Penjing Museum in Washington, D.C.

Penjing Collections

There are several major Penjing collections throughout the United States. The oldest collection is the Larz Anderson Bonsai Collection at the Arnold Arboretum in Boston, Massachusetts. Larz Anderson was an ambassador to Japan and imported bonsai in 1913. His widow donated 30 bonsai to the Arnold Arboretum in 1937 which became the Larz Anderson Bonsai Collection.

The Brooklyn Botanic Garden, in Brooklyn, New York, had an early collection of Penjing and published several handbooks

黑松 *Pinus thunbergii* 培育：斯科特·埃尔瑟
Japan Black Pine. Trained by Scott Elser

鸡爪枫 *Acer palmatum* 培育：哈维·卡拉派拉
Japanese Maple. Trained by Harvey Carapella

威廉·尼古拉斯·瓦拉瓦尼斯制作的盆景在2012年第三届美国盆栽展上展出
This Penjing was displayed in the 2012 3rd US National Bonsai Exhibition. Trained by William N. Valavanis

The Annual Forum 国际年度论坛

威廉·古拉斯·瓦拉瓦尼斯制作的盆景在2008年第一届美国国家盆栽展上展出
This Penjing was displayed in the 2008 1st US National Bonsai Exhibition. Trained by William N. Valavanis

2010年第二届美国国家盆栽展上荣获"北美珍贵树种奖"的一盆盆景
This Penjing was displayed in the 2010 2nd US National Bonsai Exhibition and received the North American Species Award

红枫 Acer rubrum 北卡罗莱纳州植物园提供的展品在2010年第二届美国国家盆栽展上展出
Red Maple displayed by the North Carolina Arboretum in the 2010 2nd US National Bonsai Exhibition

which are popular. Bonsai classes were offered to the public. Yuji Yoshimura originally came to the United States to teach and develop their bonsai collection in 1957.

The National Bonsai and Penjing Museum, located at the U.S. National Arboretum in Washington, D.C. now serves as the finest museum dedicated to promoting Penjing in the world. Classes, seasonal exhibits as well as other activities are held year around. The Museum contains a Japanese Bonsai Collection, Chinese Penjing Collection, North American Bonsai Collection as well as a Tropical Bonsai Collection.

The Pacific Rim Bonsai Collection is housed at on the Weyerhaeuser Campus in Federal Way, Washington, near Seattle. The collection includes Penjing from countries which surround the Pacific Ocean.

The Golden State Bonsai Federation in California has two significant Penjing collections. The Collection North is at Lake Merritt in Oakland, California while the Southern Collection is at the Huntington Gardens, in Pasadena, California.

Penjing Activities

A large Penjing community exists in the United States with well over 200 Penjing clubs spread throughout the vast country. Yearly Penjing conventions are held in California, Texas, Washington, Florida, New Jersey and New York.

Although the American Bonsai Society has members throughout the world, most are from the United States, Canada and Mexico. Their quarterly magazine, *Bonsai, Journal of the American Bonsai Society* includes technical articles which interest and help people to grow better Penjing.

International BONSAI is the first and only Professional Penjing magazine published in the United States and combines scholarly articles from Japan as well as from throughout the world. Now in its 36th year of publication it has set the standards for all Penjing periodical.

Most of the Penjing societies hold at least one Penjing exhibition a year to promote Penjing to the public and to enjoy their Penjing.

The interest of Penjing is alive and growing in the United States.

落羽杉 Taxodium distichum
培育：盖伊·吉德里
Bald Cypress. Trained by Guy Guidry

美洲悬铃木 Conocarpus erectus
培育：玛丽·麦迪逊
Buttonwood. Trained by Mary Madison

鸡爪枫 Acer palmatum
培育：苏鑫·苏克洛梭维斯特
Japanese Maple. Trained by Suthin Sukolosovisit

西班牙盆景一览
A View of Penjing in Spain

发言人：【西班牙】安东尼奥·帕利亚斯　图文整理：CP　　Speaker：[Spain] Antonio Payeras　Reorganizer: CP

安东尼奥·帕利亚斯 西班牙《当代盆栽》杂志执行主编
Antonio Payeras Editor & Director of *Bonsai ACTUAL* Magazine

我的名字是安东尼奥·帕利亚斯，来自西班牙。从1986年开始从事盆景事业，从此成为一名盆景专业人员。

我已经在世界各地不同的杂志上撰写了20多篇文章。我是西班牙盆景杂志《当代盆栽》的上一任编辑，现任执行主编，该杂志在全球各地发行，尤其是在西班牙、南美、欧洲和美国。我也是一名农业工程师和景观设计师。

我第一次接触盆景是在1984年瓦伦西亚市举行的花展上。由于进行上述活动，从荷兰进口了一些树，我猜想所有树木都来自中国。当我看到第一盆盆景时，我决定那将是我在未来的生活方式，所以我决定学习盆景技术。在西班牙没有人能教我如何制作一盆盆景。所以，我和其余爱好者在西班牙开始用我们当地树种，通过试验和吸取盆景养护中错误方法的经验教训来学习盆景制作。现在，30年已经过去了，西班牙已发展成为西方盆景国家中最著名的国家之一。通过当地物种，我们创造了伟大的盆景，也不再需要羡慕日本。

在西班牙主要举行两个盆景比赛。阿尔科文达斯盆栽博物馆组织的全国盆景比赛，直到2006年仍采用预选方式参与比赛时银杏奖。巴塞罗那的锦展也是一个高水准的国际盆景比赛。上述两个展会一年分别举行两次。

米斯特拉尔是西班牙最大的盆景公司，起初盆景只是公司创始人的爱好，之后他和其他亚洲盆景进出口商一起合作成立米斯特拉尔公司，在西班牙引进盆景。通过在全国范围内成功的盆景市场营销，现在几乎所有家庭都对盆景有所了解。

其他一些公司和专业人士直接从日本进口盆栽、古盆和家具，用于各大展会。他们将一些非常重要的大师的巨作从日本引进到西班牙。通过展出他们的展品，提高了我们所有盆景作品的水平，因为我们必须通过这些树木在比赛中获得胜利。

我们把盆景比作"艺术"，虽然它不像在其他国家那么普遍，但我认为西班牙的盆景代表着欧洲的最高水平。盆景在19世纪80年代通过一家名为伊贝尔盆栽贸易公司引入到西班牙。他们开始通过庞大的分销网络在西班牙引进树木，其中有一部分就在我所在的区域。这种分销网络代表着盆景在西班牙的萌芽。

几年后，因为费利佩·冈萨雷斯首相是一个狂热的盆景爱好者，盆景也开始在全国范围内流行。

目前毫无疑问的，在西班牙是路易斯·瓦列霍引领盆景的发展。他是费利佩·冈萨雷斯首相的老师，也是西班牙最成功的盆景艺术家，并且就西班牙的盆景发展而言，他对年轻一代的影响也是有很大帮助的。除了我刚才提到的路易斯·瓦列霍，西班牙最知名的盆景艺术家还有：大卫·贝纳文特、埃拉斯莫·加西亚、路易斯·维拉和加布里埃尔·罗梅罗。正是这些盆景艺术家在上届西班牙和欧洲举办的盆景大赛中赢得了大多数奖项。

西班牙是众所周知的拥有高品质山采树种的地区。也有红松树、紫杉、刺柏和我们最著名的树种——野生橄榄树。

在地中海巴利阿里群岛中最大岛屿马略卡岛和米诺卡岛上能发现地中海最好的野生橄榄树。现在这些地区的野生橄榄远销世界各地，并且在任何盆景展会上都能看到。

现在，西班牙和意大利在欧洲和西方国家领导着盆景的发展。与日本的密切关系不断丰富着我们的盆景艺术家制作盆景的技艺。另外，我认为我们与中国的关系还没达到最好的状态，在欧洲无法获取足够多的中国盆景信息。自从

The Annual Forum 国际年度论坛

《中国盆景赏石》一书的出版，我们才能知道有关中国盆景的发展趋势。

我之前对中国盆景的认识更多的是热带树木例如榕树，但现在我看到你们的盆景作品也有松柏类，这丰富了我对中国盆景的看法和观点。

此次旅行，中国盆景给我留下了深刻的印象。这是一次探索之旅，对盆景最初的起源的探索。最后，我想感谢在座的每一位听众和邀请我参加此次展览和论坛的苏放先生，谢谢大家。

My name is Antonio Payeras, I'm from Spain. I'm a Penjing professional from 1986.

I've written more than 20 articles in different magazines around the world. I'm the last editor of the Spanish Penjing magazine *Bonsai ACTUAL* which is distributed all around the world, especially in Spain, South America, Europe and USA. I'm also an agricultural engineer and landscape architect.

My first contact with Penjing was in a flower show at the city of Valencia in 1984. Some trees where imported from Holland for the event, I guess all from China. When I saw the first Penjing I decided that would be my way of life in the future. So I decided to learn Penjing techniques. No one in Spain was able to teach how to create a Penjing. So, I and the rest of the pioneers in Spain had to start with our local species and learn by trial and error method to use it in Penjing. Now after 30 years, the Spanish Penjing is one of the most prestigious in the western countries. And with the local species we have created great Penjing that nothing has to envy from Japanese bonsai.

There are two main competitions in Spain. The Bonsai Museum of Alcobendas organizes the National Bonsai Contest, which until 2006 was the pre-selection to attend the Ginkgo Awards in Belgium. Nishikiten in Barcelona is also a high caliber international Penjing competition. Both exhibitions are biannual.

Mistral is the biggest Penjing company in Spain, and together with other Asian Penjing importers were responsible for introducing Penjing as a hobby in Spain. Through the successful marketing of their Penjing throughout the country, it is now rare to find homes where they do not know about Penjing.

Some other companies and professionals are importing Penjing directly from Japan, as well as old pots, and furniture for exhibitions. Through them, some of very important masters' pieces from Japan are now in Spain, and these exhibits improve the level of all our trees, because we have to fight for the prizes in the contests with these trees.

If we talk about Penjing as an "art" it is less widespread than in other countries, but I think Penjing in Spain is at the greatest level in Europe.

Penjing came to Spain in the early 1980 through a trading company called Iberbonsai. They began to introduce trees in Spain through a distribution network, of which I was a part in my area. This distribution network was the germ of Penjing in Spain.

A few years later, and because Prime Minister Felipe González was a great Penjing enthusiast, Penjing came to be distributed nationally.

Now undoubtedly Luis Vallejo leads the Penjing way in Spain. He was the teacher of Prime Minister Felipe Gonzalez, and he is Penjing artist who has achieved the most success, and his influence with the younger generation has been instrumental in the growth of Penjing in Spain. Probably the most known artists from Spain are David Benavente, Erasmo García, Luís Vila, and Gabriel Romero, and as I said, Luis Vallejo. This bunch of Penjing artists was who achieved most of the prizes the last contests in Spain and in Europe as well.

Spain is a well know area for good quality Yamadori. Red Pines (*Pinus sylvestris*) Yew (*Taxus baccata*) Junipers (*Juniperus sabina and J. thurbinata*), and our best know species, the wild olive tree (*Olea europaea var. Sylvestris*).

The Balearic Islands, in the Mediterranean Sea, on Majorca and Minorca islands, is where the best wild olive trees in the Mediterranean can be found. Wild olives from these areas were sold around the world and now, it can be seen in any Penjing exhibition.

Now Spain and Italy are leading the Penjing in Europe and in the Western countries. The close relationship with Japan enriches continuously the techniques of our Penjing artists. On the other hand, the relation with China is not as rich as can be. There is not much information about China Penjing in Europe. Since the book of "China Penjing & Scholar's Rocks" has been published, we can see what you are doing.

My previous vision of the China Penjing were more about tropical trees like *Ficus*, but know I can see your work on Conifers and too which enriches my point of view about what you are doing.

I was very impressed by China Penjing through this trip. This was a journey of discovery and exploration on the origin of Penjing. At last, I want to thank everybody and thank you Mr. Su Fang for inviting me to the exhibition and the forum, thank everybody!

日本盆栽视窗
A Window of Japanese Bonsai

2013 唐苑的世界盆景对话
DIALOGUE TO THE WORLD PENJING

发言人:【日本】小林国雄 图文整理: CP Speaker: [Japan] Kunio Kobayashi Reorganizer: CP

小林国雄,日本水石协会理事长,春花园BONSAI美术馆馆长。
Kunio Kobayashi, the director general of Nippon Suiseki Association, the curator of Syunkaen BONSAI Museum.
小林國雄は、(社)日本水石協会理事長、『春花園BONSAI美術館』館長。

日本盆栽介绍

1. 日本盆栽的起源和发展历史

关于日本盆栽的起源,大概可以追溯到古代(奈良、平安时期)或者中世纪(镰仓、室町、安土、桃山时期)。

奈良时期(710～794)的贵族们以制作小型假山为乐。

平安时期,准确地说,盆栽更接近沙箱。但是,可以看出亲近山水的情怀在1200年前就出现了。

2. 日本盆栽界的代表人物

日本盆栽界的第一人要数小林宪雄先生,作为日本盆栽之父,受到很多人的尊敬与爱戴。小林宪雄先生认为很有必要提高盆栽艺术,决定举办日本盆栽展,历经千辛万苦,终于如愿以偿。另外,他担任盆栽推荐会的机关杂志《盆栽》的编辑。

小出信吉(第一任日本盆栽协会的理事长)继承了小林宪雄先生的精神,在距今44年前举办的大阪世博会上,将盆栽发扬光大。在1964年的东京奥运会上,举办了盆栽水石展,他们是我在盆栽界里最尊敬的人物。

3. 日本盆栽的艺术特征

日本的艺术倾向于对自然美的压缩和凝聚,而且,通过线条和空白展示余韵,使人们在闲寂与静谧中想象无形的事物。盆栽诞生于1300年前的中国,在800年前传到了日本。基于日本人的感性和美的意识,现在,盆栽的造型被确立起来,但是现在日本的盆栽在造型上一致,没有新意。我认为,只有中国具有栩栩如生线条的盆栽才属于本质上的盆景。

4. 日本盆栽作品欣赏(小林国雄藏品)

野梅 紫泥勒角底线长方盆
野梅 Prunus mume 紫泥切立下带長方鉢

梅花的魅力在于无法言喻的静默中绽放美丽,并因为四季的变换而愈发靓丽。对于盆栽来说,极为重要的一点是盆与景的协调搭配,一件作品要想达到格调高雅的艺术境界就要注意树、盆、几架三位一体的和谐

The charm of plum blossom is the beautiful blooms in the unspeakable silence, and it becomes even more beautiful because of the change of seasons. For the bonsai, it is extremely important for the pot to harmonize and match with the scene. For an artwork, in order to reach the artistic state of elegant style, the attention shall be focused on the harmony of three factors as the tree, the pot and the shelf.

梅花は、魅力が言えない静けさに美を現すことにあり、四季の変化することにてより美しくなる。盆栽に対して言えば、大切なのは、盆と景が調和的に組み合わせるということである。ある作品を上品な芸術境地に達させていく場合、樹、盆、卓を一体にする調和を重んじることは必要である。

The Annual Forum 国际年度论坛

深山海棠 广东切角长方盆
深山海棠 *Malus sieboldii* 広東隅切長方鉢

木瓜（东洋锦）古渡白交趾飘口长方盆
木瓜（東洋錦）*Chaenomeles lagenaria* 古渡白交趾外縁隅入長方鉢

唐棣 古渡紫泥菱花圈足盆
采振木 *Amelanchier asiatica* 古渡烏泥輪花鉢
唐棣之美在于其具有荒皮性，在日本有的人在满开的唐棣树下载歌载舞、享受生活。此盆栽的配盆来自中国，在日本并没有此种类型的盆。
The beauty of shadbush is in that it has a rough surface from nature. In Japan, people would dance and sing under the blooming shadbushes to enjoy their lives. The pot matched for this bonsai is from China, there is not such kind of pot in Japan.
采振木の美は、おそらく荒皮性なのであろう。日本において、良く咲いている采振木の下で歌ったり踊ったりして楽しい時間を過ごす人たちもいる。また、この鉢は中国製のもので、日本ではこのような鉢はない。

长寿梅附石盆栽 均釉袋式椭圆盆
長寿梅石附 *Chaenomeles japonica* 均釉袋式楕円鉢

花梨 古渡广东切角长方盆
花梨 *Pseudocydonia sinensis* 古渡広東隅切長方鉢
此盆栽的配盆非常考究，其雕工、质地都极好，仔细观赏，颇令人感动。
The pot matched for this bonsai is very exquisite. The carving technique and quality is excellent. I feel very touched when I carefully appreciate it.
この鉢はかなり工夫され、その彫刻も材料も非常に優れる。よく観賞すれば、感動させられるようになる。

"圣山"刺柏 古渡乌泥抚角长方盆
获第25届日本盆栽作风展"环境厅厅长奖""聖山" 杜松 *Juniperus rigida* 古渡烏泥撫角長方鉢 第25回日本盆栽作風展「環境庁長官賞」
此盆栽最初的时候高达1.5m以上，后来经拦腰截断，重新制作整理，现只有0.5m。对于盆栽来说，取其精华、盈缩成寸是最重要的。
At the first, this bonsai was above 1.5 m high, and later was cut off in the middle for recreation and trimming. It is now only 0.5 m high. For bonsai, it is most important to extract the essence and shrink the size.
この盆栽は元々高さが1.5m以上、その後、真ん中で切られ、再び制作 整理され、いま、僅か0.5mとなる。盆栽に対して言えば、その優れる分を取り出して小さくさせるのは最も重要なことである。

"奥之巨松" 五针松 南蛮外边长方盆
「奥の巨松」五葉松 *Pinus parviflora* 南蛮外縁長方鉢
在小林先生的一本书中，首页便是这盆盆栽。小林先生说，他在28岁的时候第一次见到这盆盆栽，也是在28岁的时候第一次接触盆栽并爱上盆栽。正是这盆盆栽展现出的顽强与坚韧的生命箴言，让他陷入对盆栽的爱慕。
This bonsai is on the first page of one of Mr. Kobayashi's books. Mr. Kobayashi said that, he has first saw this bonsai when he was 28, and it was also the first time he has been attached to bonsai and in love with it. The tenacious and tough life proverbs expressed by this bonsai has made him fall in love with bonsai.
小林さんの著書の表紙に、この盆栽を載せている。28歳の時、この盆栽を始めて見て、始めて盆栽に強い趣味を持ったと言った。この盆栽に現された粘り強さ及び強靭な生命のコンセプトにて、彼は盆栽が大好きになった。

2013 唐苑的世界盆景对话
DIALOGUE TO THE WORLD PENJING

"黑龙" 黑松 紫泥外边长方盆 获第71届国风盆栽展"国风奖"
「黒龍」黒松 *Pinus thunbergii* 紫泥外縁長方鉢 ※第71回国風盆栽展「国風賞」

"国之镇" 鱼鳞松 紫泥袋式椭圆盆
「国の鎮」蝦夷松 *Picea glehnii* 紫泥袋式楕円鉢

此盆栽的造型为"迎宾式",好似在欢迎客人。在中国的盆景展场上和街道两旁我看到过很多这种迎宾造型的树。
The modeling of this bonsai is as the gesture of "welcoming", which looks like to welcome the guests. In the bonsai exhibition and at both sides of streets in China, I have seen many trees in this modeling.
この盆栽の姿は「歓迎型」となり、まるでお客を迎えているようである。中国の盆景展示場及び街道の両側で、このような姿の樹を多く見たことがある。

山枫 均釉椭圆盆 获第72届国风盆栽展"国风奖"
山モミジ (*Acer palmatum*) 均釉楕円鉢 第72回国風盆栽展「国風賞」

30岁的女人既有20岁女子的活力娇美,又有40岁女人的成熟智慧,是女人一生之中最美的时刻。山枫也是如此,秋季的山枫既有夏日的明媚又有冬日的成熟,这盆盆栽恰是在最美的季节拍摄的。另外,对于盆景来说线条是展现美的关键,这盆山枫就以优美的线条展现了其窈窕的"身材"。

Women in their 30's have the vigorous charming of 20-year-old women and the matured wisdom of the 40-year-old. It is the most beautiful period in a woman's life. And so the mountain maple is. The mountain maple in autumn has the brightness of summer and the maturity of winter, and this picture of bonsai was taken in the most beautiful season. Additionally, for bonsai, the lines are the key to express the beauty. The exquisite lines of this mountain maple have revealed her gentle and graceful "body".

女性に例えると、30歳の女性は20歳女性の活気・若さ・美しさを有するうえ、40歳女性の成熟・知恵をも有するので、女性の一生の中に最も美しい時期と思う。山モミジもそうである。秋の山モミジは夏日の明るさを備えるうえ、冬の成熟を備え、この盆栽はちょうど最も美しい季節で撮られたものである。そのうえ、盆栽に対して言えば、ラインは美を現すキーポイントであり、この山モミジは美しいラインを利用して窈窕たる姿を現している。

5. 日本盆栽展览会
日本最受瞩目的盆栽展是国风盆栽展,历史悠久,审核严格,展示着高品质的盆栽。

6. 日本盆栽媒体
盆栽杂志有《近代盆栽》,水石界有《爱石》,皋月协会有《杜鹃研究》。

中国盆景印象

我感受到中国盆景线条的作用和"气"是很享受的事情,但是,感受简约与空白形成的余韵一样让人向往。另外,如果配置桌几和青苔等来提高意境的话,则效果最好。

世界盆栽的发展意见

"盆栽·BONSAI成为世界性语言"为了将盆栽全球化,有必要召开类似奥运会那样的大会,举办世界盆栽爱好者聚会的活动和交流。另外,不只是在技术层面,在精神层面上,应该将"盆栽道"发扬光大。

"云龙" 真柏 和钵外边长方盆 "松树千年翠"挂轴
「雲龍」真柏 (*Juniperus chinensis* var. *sargentii*) 和鉢外縁長方鉢、掛軸「松樹千年翠」

小林先生曾和须藤雨伯先生一同拜师片山先生,并在片山先生的指导下学习景道3年。此盆栽树龄在百年以上,摆放在小林先生家的壁龛中,其摆放与布置是典型的日式陈列风格,树、盆、几架、配饰等都经过仔细琢磨与研究,具有"景道"理念与意境。为了展示与欣赏盆栽和水石,小林先生在他的春花园BONSAI美术馆设有这样的壁龛10余个。

Mr. Kobayashi and Mr. Uhaku Sudo have been apprenticed to Mr. Katayama and learned Keido for 3 years under Mr. Katayama's mentoring. The tree age of this bonsai is above a hundred years. In Mr. Kobayashi's house, the bonsai is placed in the tabernacle. The layout and arrangement is in the typical Japanese display style. The tree, the pot, the shelf and decorations have been carefully considered and researched and containing the concept and artistic conception of "Keido". In order to display and appreciate bonsai and suiseki, Mr. Kobayashi has set more than 10 tabernacles in his Shunka-en BONSAI Museum.

小林さんと須藤雨伯さんは一緒に片山一雨さんを先生にして学び、且つ片山先生の指導の元で景道を3年学んだことがある。この盆栽は樹齢が百年以上、小林さんの家の床の間に置かれ、その配置は典型的な和風陳列風格となり、樹、盆、卓、飾りなどに細かく工夫して研究したことがあり、「景道」のコンセプトと意境を備えている。盆栽と水石を展示・鑑賞するために、小林さんの春花園BONSAI美術館においてこのような床の間は10室以上も設けている。

The Annual Forum 国际年度论坛

Japan

1. Origin and development history of the Japanese bonsai

The Japanese bonsai origin can be traced back to the ancient times (Nara Period and Heian Period) or the Middle Ages (Kamakura Period, Muromachi Period, and Azuchi-Momoyama Period).

The nobles in Nara Period (710~794) make the small rockwork for fun.

Accurately speaking, the bonsai in Heian Period is much close to the sandbox. However, the feeling getting close to the landscape appeared 1200 years ago.

2. The representative individual of the Japan bonsai field

The most important person of the Japanese bonsai industry is Mr. Norio Kobayashi, as the Father of Japanese Bonsai, he is respected and loved by many people. Mr. Norio Kobayashi thinks it is necessary to improve the bonsai technology and determines to hold the Japanese bonsai exhibition. And, after having experienced innumerable trials and hardships, he has achieved what he wishes eventually. In addition, he holds the post of the editor of the organ magazine-Bonsai of the Bonsai Recommendation Association.

Shinkichi Koide (the first President of the Japanese Bonsai Association) inherits the spirit of Mr. Norio Kobayashi and carries forward the bonsai in Osaka World Exposition 44 years ago. In 1964 Tokyo Olympic Games, the bonsai and water stone exhibition was held. And, they are the figures I most respect in the bonsai industry.

3. The characteristics of the Japanese bonsai

The Japanese art is inclined to the compression and cohesion of the natural beauty, which applies lines and blanks for displaying the lingering charm to make the people imagine intangible things. The bonsai originated in China 1,300 years ago, which was spread to Japan 800 years ago. Based on the Japanese's sensibility and beauty sense, the bonsai shaping has been established currently. However, current Japanese bonsais have consistent shaping, and there is no any new meaning. I think, only the Chinese bonsai with vivid lines can be classified as the bonsai essentially.

4. Pictures of Mr. Kobayashi's bonsais

5. The most famous bonsai exhibition in Japan

It is the National Style Bonsai Exhibition, which has long history and strict review and can represent the high-quality bonsai.

6. The best bonsai media (book/magazine/television program)

The bonsai magazine covers the Modern Bonsai; the water stone industry includes the Love Stone, and Gaoyue Association has the Gaoyue Monthly Magazine.

China

It is enjoyable to feel the line effect and "spirit" of Chinese Penjing; however, it is greatly admired to feel the lingering charm formed by simplicity and blank. In addition, if preparing stools and moss, it will bring the best effect.

World

"BONSAI" has become a world language; to globalize the bonsai, it is necessary to hold meetings like the Olympic Games and activities and exchanges for the worldwide bonsai fans. In addition, it is recommended to carry forward the "bonsai path" on the technical and spiritual levels.

日本盆栽の窓

演説：小林國雄　写真/文章整理：CP

日本について

1. 日本盆栽の起源と発展歴史

日本盆栽の起源は古代（奈良・平安時代）や中世（鎌倉・室町・安土・桃山時代）が盆栽前史といえるかもしれない。

奈良時代（710～794年）の貴族達は小さな仮山（仮の山）を作って楽しんでいた。

平安時代には盆栽というよりは、箱庭のような盆景に近いものだった。しかし山水景を身近に楽しもうとする精神は1200年以上も前に発露されていたのである。

2. 日本盆栽界の代表的な人物

日本の盆栽界における一番の恩人は何と言っても小林憲雄先生である。盆栽界育ての親として多くの人に尊敬されている。小林氏は芸術としての盆栽の認識を高める必要を痛感し、国風盆栽展を開催する決心をし、あらゆる苦心の末についに実現したのである。また盆栽推奨会の機関誌「盆栽」の編集にあたった。

そして、小出信吉（初代日本盆栽協会理事長）は小林憲雄先生の志を受け継ぎ、今から44年前に催された大阪の万国博覧会で盆栽を隆盛にした人物である。また、1964年東京オリンピックでも盆栽水石展を開催し、盆栽界で私が最も尊敬する人たちである。

3. 芸術から見ると、日本盆栽の特徴

日本の芸術は自然美を圧縮・凝集するように思われる。そして線と余白によって余韻をもたせ、侘び寂びの中から目に見えない物を想像させることにある。盆栽は1300年前に中国で生まれ800年前に日本に伝わってきた。日本人の感性と美意識によって今の日本の盆景の型が確立されたが、今の日本の盆栽は型にはめ込まれた画一的になっていてつまらない。私は中国の躍動感のある、線を生かした盆栽こそ本物であると感じている。

4. 小林先生の盆栽の中で12枚用意した

5. 日本全国で一番の盆栽メディア

日本の盆栽雑誌では「近代盆栽」であり、水石界では「愛石」また皐月においては「さつき研究」であると思う。

6. 日本最高の盆栽メディア

盆栽展示会　国風盆栽展である。やはり歴史が長いことや審査による選択がありことで質の高い盆栽が陳列される。

中国について

中国盆景の線の動きと「気」を感じさせる事は素晴らしいと思う。しかし、簡略することや余白による余韻を感じさせる事にも目を向けて欲しい。また、卓合わせ、下草などにも気を配られたら最高である。

世界について

「盆栽・BONSAI」は今世界語となっている。盆栽をグローバルにするためには、オリンピックのような大会を開き、世界の盆栽愛好家が集うイベントや話し合いの場を催すべきであると考える。また、技術面だけでなく盆栽道という精神面での高揚に目を向けるべきである。

2013 唐苑的世界盆景对话
DIALOGUE TO THE WORLD PENJING

论及历史与未来 谈说理念与议题
——从盆栽与水石两方面说起

Discussed the History and Future, Talk about the Concept and Doubts-- About Bonsai & Suiseki

发言人：【日本】须藤雨伯 图文整理：CP　Speaker：[Japan]Uhaku Sudo Reorganizer: CP

须藤雨伯，景道家元二世，竹枫园园主。
Uhaku Sudo, Keido Lemoto (headmaster) II, the owner of Chikufuen.
須藤雨伯は、景道家元二世、竹楓園園主。

今天能在此与在座的各位盆景友人探讨我感到荣幸之至。首先，就小林国雄先生刚才提到的景道，我稍作说明。我和小林先生曾一同拜师景道的片山一流老师，在学习盆栽后的第二阶段我们学习了景道。以我的观点来看，景道指引着日本盆栽之路，告诉人们盆栽是什么、水石是什么，并告诉人们如何通过盆栽美学、灵魂、日本文化等进行盆栽装饰和摆放。我认为景道的"景"即是指盆景的"景"。日本盆栽之路指引着很多人，因此，通过盆景将获得怎样的人格是我现下探索的。

日本盆栽与世界盆景

1. 日本盆栽和世界盆栽的关系

很多盆栽友人经常到大宫盆栽村（现在的盆栽市）的九霞园——村田久造先生的住所拜访，其中包括日本盆栽人，也包括美国、法国、德国等世界各地的盆栽友人。九霞园的村田先生精通英语，可以向很多外国人介绍盆栽。尤其是村田先生与当时的首相吉田茂是朋友，并担任吉田茂的盆栽指导人和管理负责人。因此，他能结识到日本的许多代表性政治家，属于建立盆栽文化发展基础的人物。同时，他也管理日本天皇的盆栽，将日本盆栽推向了全世界。

村田先生为了向美国宣传盆栽，付出了巨大的努力。而且，他介绍日本的吉村香风园的继承人吉村先生作为美国盆栽的指导人。之后，吉村先生移居美国，开办了盆栽学校，为大量在美国的日本人和白人宣传真正的盆栽。在美国，为了盆栽文化的发展，他将一生的精力浓缩在盆栽的宣传上。

在他的学生中，有一位叫做约翰·纳卡（John Naka）的日本籍人，约翰·纳卡开创了美国盆栽的先河，被尊称为美国盆栽之父，在美国全国培养了许多盆栽人。而且，他奠定了世界盆栽的基础，积极活动，成为为世界培养了很多盆栽指导者的原动力。

1970年，我受村田久造先生的叮嘱，拜会了约翰·纳卡，与他一起在美国的东西部进行巡回演讲。这是我最早对美国盆栽进行指导，之后，数次受他之邀去美国进行指导。1989年我带木村正彦先生去了美国，并且策划了BCI在美国的盆栽现场制作表演，约翰·纳卡担任翻译，表演受到了很多盆栽爱好人的青睐，也为现在美国盆栽的发展奠定了基础。

2. 日本盆栽为世界盆栽带来的影响及其根本原因

可以认为，村田久造、吉村香风园、约翰·纳卡的功绩在于BCI的成立和发展。受他们的影响，很多日本盆栽作家在美国进修、演讲，美国盆栽的发展与世界盆栽的发展紧紧地联系在一起。特别是约翰·纳卡的贡献非常大，可以说如果离开约翰·纳卡，现代世界盆栽的发展将无从谈起。

日本盆栽之美、艺术性、日本文化之美感动着世界上很多人。日本盆栽美学、闲寂、幽静的美学是美妙绝伦的主题。

3. 近年来，日本盆栽界在世界盆栽界中的发展

"日本的盆栽是世界盆栽的主流"并非永不改变。中国盆景的发展将是世界盆栽取得更大发展的巨大原动力，盆栽文化的全球化已经形成，将取得更大发展。

中国盆景对世界盆栽的影响非常大，中国盆景应该能改变世界盆栽的未来。中国盆景的历史、思想、宗教、哲学、美学的理论和合理性俱佳，全世界的人易于理解。我相信，中国盆景的发展将影响日本盆栽的发展，进而影响世界盆栽的发展。

中国盆景的相关议题：
①盆栽与道教的关系？
②在天人合一中人应该做什么？

③意境是什么?
④山水思想是什么?
⑤盆景的产生目的(秦始皇的盆栽与盆景)?
⑥庭园理念与盆景理念的关系?
⑦为什么盆景存在了2000多年?
⑧是什么原因,让皇帝特别钟情于盆景?
⑨为什么清朝乾隆帝特别关心盆景并制作了精美的盆器?
⑩这个时期(清乾隆时期)盆景发展状况是什么?

日本盆栽向中国学习的东西很多,但是,却没有可以教给中国的。原因在于日本盆栽是将中国盆景放在日本土壤中培养而成,并获得了巨大成功,从本质上说,什么都没有改变。道教、儒教、佛教(禅)是日本盆栽的根本理念,与中国盆景的理念一致。中国的神仙思想、长生不老的理念与日本的禅的自然思想、闲寂、幽静的美学是相通的。

水石之美

以下是我引用前人的语言对水石之美进行概括。如果能得到大家的认可将使我荣幸之极!因为这是我的水石理念。

水石蕴涵山水和情景,可以欣赏自然美。
山水不是单独的山脉或河流,而是与人世间相对的精神象征,与哲学、宗教、历史、文学、美术等文化价值密不可分。从本质上说,水石与山水画或假山(山水庭园、枯山水)基于相同的目的被人们所感受,将山水表面化以后就成了山水画,立体化以后就成了假山,放置于盆上就成了盆石(水石)。水石是立体的山水画,被称为无声的山水诗。此时,山水与单独的自然风景不同,具有非常广泛的概念,不仅仅包括自然的风景(真山水),也包括胸中的山水、自然的风景山水、宗教的山水(神仙、道教、蓬莱山、佛教的须弥山)等。

在中国,对于山水的理念已经形成成熟体系,是精神性极高的概念。孔子曾经说过:"智者乐水,仁者乐山"(《论语》)。六朝时期的宗炳(375～443)曾经说过:"至于山水,质有而灵趣"、"山水以形媚道"。日本的道元禅师曾经说过:"近处山水是现成的古佛之道"(《正法眼藏》)。因此,山水画、盆石、假山(山水庭园)与山水的诗文一样,追求高深的精神。

"山水"属于盆石(水石)的哲学范畴,是水石存在的根本。
山水画是描写山水的艺术,本来的目的在于彰显"气",所以,气韵栩栩如生。如果把伟大、灵秀的山水之气用画表现出来,观者会与山水之气成为一体,可以徜徉在山水的世界里,这就是所谓的"神游"。

山水之气在于石,也称为"气的核心"。在中国,名山、大川的石头受到人们的珍爱,是因为人们喜欢享受灵山不可思议的原始的气。中国的灵璧石作为赏石(水石)被喜爱者放在身边赏玩也属于相同的理由。

中国的这些神仙道教思想即是日本文化的思想,水石亦是由此得以成立。

盆栽和水石作为日本人热爱自然的根本,具有以传统方式孕育的深奥趣味。

"水石"是山水景石的简略语。水石的定义是:"自然地表现一个自然景观之美"。另外,大贯忠三曾经写过:"石表现风景,水增加变化。通过动与静、硬与柔,产生的境界显示了极高的水石之美。"赖山阳曾经写到:"非常喜欢自然风景,渴望游览大自然的美妙境界,仿佛羽化登仙一样。"

水石由大自然中伟大的力量产生,在自然之中发现美是指以每个人的艺术意识为根本。为了提高该意识,增加对石头的了解,通过静静地观察美的事物,在任何地方都追求真正的美——这是高水平观赏的一个条件。

石像镜子一样,反射出我的模样,这种对深奥境界的追求正是爱石兴趣的精髓——宫坂隆知曾经这样写过。

幽玄,主要指淡薄深邃的心灵意境,铺开让人"感动"的画面,展示出令人回味之美称为幽玄。

人们认为只有幽玄才是中世纪美意识的中心,是形成了理念的东西。分析幽玄美的结构,探求其功能,结果非常"闲

Discussed the History and Future Talk about the Concept and Doubts — About Bonsai & Suiseki

Speaker: Uhaku Sudo Reorganizer: CP

寂"、"幽静",根据"幽玄是未被表现的回味"的观点,显示暗示性,而且,该暗示性是真实的,表现的效果成立。

形成幽玄的结构与"语言"、"形态"、"心"有关。

人类与"物"一体产生了美,即完成了美的态度。所以,作为宇宙中存在的原理,对无常的自我意识是美的原理,是美的态度的根本。

由世界盆景石文化协会主办的国际论坛议题(关于水石的理解与质疑):

①日本的水石是根据石头的形态,想象山水的情景,并享受其中的乐趣。水石是指"山水"的水、"情景石"的石。

②中国石文化的分类方法和历史?
中华民族对石的基本理念是什么?
石的精神和美的意识?

③石的美学
在山水之美以外,追求什么美?

④"人类生于石并回归于石"的说法存在吗?
会回到赏石精神的根本吗?

在日本,盆栽与水石被称为车之两轮,我想是因为两者在理念上是通的。

中国盆景与石文化的关系是什么?共同点又是什么?

⑤关于日本水石与中国赏石的共性、美学、思想、哲学等。

⑥关于五代时期的赏石文化、六朝时期宗炳的山水思想与赏石理念。

I feel honored to discuss with all of you bonsai friends present here today. First, I'll make brief description in terms of keido as Kunio Kobayashi mentioned. Kunio and I have ever taken Ichiu Katayama who is a master of keido as our teacher. We've studied keido in the second stage after learning bonsai. In my opinion, keido guide Japanese bonsai's direction, tell people what bonsai is, what suiseki is and tell people how decorate and display bonsai through bonsai aesthetics, soul and Japanese culture. I think the "Jing" of keido is namely the "Jing" of Penjing. Japanese bonsai guided the way for many people. Therefore, how to access the character through Penjing is my exploration now.

Japanese Bonsai and World Penjing

(1) Relation between Japanese bonsai and world Penjing

Japanese bonsai, world bonsai (USA, France, German, etc.), Kyuka-en of Omiya bonsai village (current bonsai city), and the residence of Mr. Kyuzo Murada are frequently visited by many foreigners. Mr. Kyuzo Murada of Kyuka-en is perfect in English, so he can introduce the bonsai to many foreigners. Especially, Mr. Kyuzo Murada is a friend of the former prime minister-Shigeru Yoshida, who is the bonsai trainer and management principal of Shigeru Yoshida. Therefore, as he can get to know many Japanese politicians, he is the man to establish the bonsai culture development foundation. And, he also takes charge of management of Mikado's bonsai meanwhile introduces Japanese bonsai to the world.

Mr. Kyuzo Murada has made great efforts to introduce the bonsai to USA. And, Mr. Yoshimura, the inheritor introducing the Japan Yoshimura Xiangfeng Garden as well as the USA bonsai instructor, has moved to USA and set up a bonsai school to propagandize the true bonsai to a number of Japanese and white races in USA. And, he has spent his whole life for the development of bonsai culture in USA.

Among his students, there is one Japanese named John Naka who has initiated the USA bonsai and is regarded as the Father of USA Bonsai. In addition, he has trained a lot of bonsai people in USA; established the foundation of the world bonsai and actively took part in all kinds of activities, thus became the original motive power to train a lot of bonsai directors.

In 1970, requested by Kyuzo Murada, I paid a courtesy call to John Naka and made tour lectures in the Eastern and Western parts of USA. This is my first guidance to USA bonsai,; afterwards, I have been invited by him to give guidance in USA for several times. In 1989, I took Masahiko Kimura to USA and planned the BCI bonsai live demonstration in the USA. With John Naka serving as the interpretation, the demonstration has been favored and admired by lots of bonsai fans and laid a solid foundation for the development of USA bonsai.

(2) What effect does Japanese bonsai bring to the world bonsai? What is the root cause?

It can be thought that the achievements of Kyuzo Murada, Yoshimura Kofuen and John Naka lie in the establishment and development of Bonsai Club International. Affected by them, many Japanese bonsai writers have studied and given lectures in USA, and the development of USA bonsai is closely linked together with the development of the world bonsai. Especially, John Naka has contributed a lot.

In a manner of speaking, there will be not the development of modern world bonsai without John Naka. And, the beauty and artistry of Japanese bonsai have moved a lot of people in the world. The aesthetics of Japanese bonsai and the carefree and peaceful aesthetics are magnificent topics.

(3) In recent years, how does Japanese bonsai community judge the

development of world bonsai community?

The principle that "Japanese bonsai is the mainstream of the world bonsai" may change. However, the development of the China Penjing will be a huge motive power for the greater development of the world bonsai. The globalization of bonsai culture has been formed, which will achieve greater development.

The impact of the China Penjing on the world bonsai is very large, and China Penjing can change the future of the world bonsai. China Penjing"s history, ideology, religion, philosophy, aesthetics theory and rationality are perfect, which are easy for the people around the world to understand. I believe that the development of China Penjing will impact the development of Japanese bonsai, thus affecting the development of the world bonsai.

Questions about China Penjing:
① Relation between the bonsai and Taoism
② What should one do in the "theory that man is an integral part of nature"?
③ What is the artistic conception?
④ What is the landscape ideas?
⑤ What is the purpose to produce Penjing (bonsai and Penjing of the Emperor Qin Shi Huang)
⑥ What is the relation between the garden concept and Penjing concept?
⑦ Why can the Penjing have a history of more than 2,000 years?
Why does the emperor have special preference on Penjing?
Why does the Emperor Qianlong of Qing Dynasty pay particular attention to Penjing and make the exquisite Penjing pot?
How about the Penjing development conditions in the period of the Emperor Qianlong?

Japanese bonsai can learn a lot from China, but nothing can teach to China.Japanese bonsai is made by placing China Penjing into Japanese soil, which gains huge success. In essence, nothing has changed. Taoism, Confucianism, Taoism, Confucianism and Buddhism (Chan) are the basic concepts of Japanese bonsai, which are consistent with that of China Penjing. Chinese Taoist thoughts of immortals, the concept of immortality and Japanese Chan's natural thought and carefree and peaceful aesthetics are interlinked.

About Suiseki
The following is the summary of suiseki beauty with the predecessors' words.

I will be most honored if you all agree with me, for it is my water stone concept.

About suiseki beauty:
Suiseki contains the landscape and the scene, so that we can appreciate the natural beauty.

The landscape doesn't refer to the separate mountain or river, but the spiritual symbol compared with the secular world, which is closely related to the philosophy, religion, history, literature, art and other culture values. Essentially, suiseki and landscape painting or rockwork (landscape garden and rock garden) are perceived by people with the same purpose, which can be turned to the landscape painting by surfacing the landscape; the rockwork by three-dimensional treatment, and the potted stone (water stone) by placing on the pot. suiseki is a kind of three-dimensional landscape painting, which is called the silent landscape poem. In such case, the landscape is different from the separate natural scene, which has broader concept, containing the natural scene (true landscape), landscape in heart, natural scene and landscape, religious landscape (Taoist immortal, Taoism, Penglai Mountain, and Buddhism Sumeru Mountain), etc.

The landscape concept in China has been shaped into a matured system, which is a kind of concept with higher spirituality.

Confucius ever said, "Wise man favours water, benevolent man favours mountains" (Analects of Confucius).

Zong Bing (a famous landscape painting critic in Six Dynasties, 375~443) ever said, "Landscape, 'the quality of the spirit and interest'" and "Truth revealed in mountains and waters". Dogen from Japan ever said, "The nearby landscape is the ready-made Buddhist path" (Correct Dharma and Eye-treasury). Therefore, the landscape painting, potted stone and rockwork (landscape garden) are same as the landscape poems and essays, pursuing for the profound spirit.

The fact that the "landscape" is categorized into philosophy of the potted stone (water stone) is the basis for existence.

The landscape painting is a kind of art describing the landscape, and its original purpose is for "spirit" transition. Accordingly, the artistic conception is natural as though it were living.

If representing the great and exquisite landscape spirit with paintings, the audience can integrate with such landscape spirit and stroll in the landscape world. And, this is the so-called "inner journey".

The landscape spirit lies in the stone, which is called the "core of spirit". In China, the fact why stones of famous mountains and great rivers are cherished by people is that they like enjoying the miraculous original spirit from the soul mountain. And, the stone fans cherish Chinese Lingbi stone and make it as the stone appreciation (water stone) for the same reason.

With chinese Taoism though becomes as Japanese culture thought, suiseki has gained acceptance.

As the fundamental for Japanese to love the nature, the bonsai and water stone have the traditionally profound interest.

The "water stone" is the abbreviation of the landscape stone.

Suiseki is defined as "a beauty naturally representing a natural scene".

And, Tyuzan oduki said, "Stone can represent the scene and water can bring more changes. Incomparable water stone beauty can be represented with the dynamic and the static, the hard and the soft, and the generated realm.

And, Sanyo Rai wrote: "I like natural scene very much and I'm eager to visit the beautiful nature, just like taking flight to the land of the immortal."

Suiseki comes from the great power of the nature. Finding the beauty in the nature means taking everyone's art consciousness as the root. To improve such consciousness, increase understanding of the stone, quietly observe someone or something beautiful, and pursue for the true beauty anywhere----this is the premise for high level appreciation.

The stone is like a mirror, which can reflect my look; the pursuit for the profound realm is the essence of the stone loving----Takatomo Miyasaka wrote these words.

Tacit (mainly referring to indifferent and profound soul)
The tacit means paving the "touching" picture and showing the evocative beauty.

It is believed that the tacit is the center of the beauty consciousness of Middle Ages, which forms the idea. After analyzing the structure of the tacit beauty and seeking its functions, the result is very "carefree" and "peaceful". According to the viewpoint of "the tacit is a hidden aftertaste", we know the tacit is suggestive, and its suggestibility is true, which has established performance effect.

The structure of the tacit is related to the "language", "shape" and "heart".

Once the human being and the "object" generate the beauty, the aesthetic attitude is completed. Accordingly, as a principle for existence in the universe, the uncertain self-awareness is the beauty principle and the root for the aesthetic attitude.

Understanding and question of suiseki:

① For Japanese suiseki, one can imagine the landscape scene and enjoy the fun herein based on their shapes. suiseki contains the water of the "landscape" and the stone of the "scene stone".

② Classification method and history of Chinese stone culture

What is the basic concept of stone for Chinese nation?

Spirit of stone and sense of beauty.

③ Aesthetics of stone

Besides the landscape beauty, any other beauty would be pursued for?

④ Does the statement "the human being is born from stone and return to stone" exist?

Can the human being return to the essential of the stone appreciation spirit?

In Japan, the bonsai and suiseki are called two wheels of vehicle, and I think it is because they share the same communication concept.

How about the relationship between chinese Penjing and the stone culture? How about the common point?

⑤ About the generality, aesthetics, thought and philosophy between Japanese water stone and Chinese stone appreciation.

⑥ About the stone appreciation culture of the Five Dynasties and landscape thought and stone appreciation concept of Zong Bing (a famous landscape painting critic in Six Dynasties).

世界盆景石文化協会主催の盆景論壇活動議題

㊀日本の水石は山水景情を石の形姿より想像して楽しむもので、水石とは山水の水と景情石の石をもって水石としておる

② 中国石文化の分類のあり方と歴史 中国民族が石に求めてきた基本理念とは何か

③ 石の美学 山水美以外にどのような美を追求しているのか

④ 人間は石から生まれて石へ帰るの言葉は存在するか

⑤ 日本の水石と中国の賞石の共通性、美学、思想、哲学等について

⑥ 五代時代の賞石文化、六朝時代の宗炳の山水思想と賞石の概念について

石が鏡のように私の姿を写している、この深い境地の追求こそが愛石趣味の神髄と宮坂隆知氏は書いている。

「あわれ」の極まる所に開けてくる余情美を幽玄と呼ぶ。

幽玄こそ中世美意識の中心、理念をなすものと考えられる。幽玄美の構造を分析し、その様式の本領を探っていくと結局「わび」「さび」が究極であり、幽玄は表現されない余情であるという点で暗示性を立ててしかもその暗示性は真実で的確で表現の効果であり、美的態度としての無常の自覚が美の原理における存在の原理としての無常の自覚が美の原理であり、美的態度の根本基底であるとしているのである。

その幽玄を成立させる構造を「ことば」と「姿」「心」と「物」との一体化に美が成立し、美的態度の完成があるとしているのである。従って宇宙におけることばとしている。

山水画は山水を写す芸術であるが本来の狙いは「気」を移す事である。それによって気の韻（ひびき）が生動するのである。

至高霊妙なる山水の気を画図に写す事ができれば観者は山水の気と一体となり、山水の世界に逍遥する事ができる。「これがいわゆる「神遊」である。

山水の気そのものが石であり「気の核」とも言われている。中国では名山、大川の石が珍重されるが、それはそれらが霊山の霊妙な気をそのまま享けているからである。

盆石（水石）中国の霊璧石を身近に置いて楽しむのも同じ理由である。

これらの中国の神仙道教上の思想が日本の文化達の思想となり、水石が成立してくる。盆栽も水石も日本人の自然愛の心の根底として伝統的に育まれた奥ゆかしい趣味である。

水石とは山水景石の略語である。

水石の定義として「無理をしないでもてる程度の大きさの一個の自然的で自然景観の美を表現しているもの」

そして大貫忠三氏は、石が景勝を表現し水がこれに変化を添える。この動と静、硬と和によってかもしだされる境地こそが水石美の極致と書いておる。

又頼山陽は風物自然を擬し、また大自然と遊ぶ神韻渺望たる境地、または自ら羽化登仙するのであろうと書いている。

大自然の偉大なる力によって創成されたものであり、その自然の中に美を発見するということは、各個人の芸術意識を根底とする働きであり、その意識を高める為には石についての知識を深める事、美しいと感じる物を静かに眺める事によって美の正体をどこまで追求する、と言うこの高度な観賞をする事が一条件となる。

日本の道元禅師は「而近の山水は古仏の道現成なり」と言っている《正法眼蔵》従って、山水画・盆石・仮山（山水庭園）は山水の詩文と同様、高く深い精神性を宿す事が求められる。

「山水」は盆石（水石）の哲学であり存在根拠である。

質は有にして趣は霊なり」「山水は形を以て道は質に有にして趣を霊なりにす」と言っている。

歴史と未来を議論 理念と議題を相談——盆栽と水石の両方から語る

演説：須藤雨伯　写真／文章整理：CP

皆さん、御慶栄耀。今度の論壇に参加することに、たいへん嬉しい。どうもありがとう。先ほど、小林先生が論説した景道にてちょっとご紹介いただく。私が小林先生と一緒に景道を勉強した。私の今の立場から見ると、景道を第二段階として勉強している。私の今の立片山一流先生に第二段階として景道を勉強している。私の今の立場は何か、そして、日本盆栽の道を教えている。盆栽とは何か、水石とは何か、盆栽の美学観・霊魂・日本文化から盆栽の飾り方と作法を教える。景道の景は盆景の景である。日本盆栽の道が多くの人に教えて、そして、盆景を通して人格がどういうことかが私が探している。

日本盆栽と世界盆栽

（1）日本盆栽と世界盆栽との関係

日本盆栽と世界の盆栽（アメリカ・フランス・ドイツ等）、これらの国から、大宮盆栽村（現在の盆栽町）の九霞園、村田久造氏の所へ多くの外国人が訪問している。九霞園の村田先生は英語が堪能でしたので、多くの外国人に盆栽を紹介出来た。特に村田先生は当時の総理大臣、吉田茂と交友があり、吉田茂氏の盆栽の指導者であり、管理責任者であった。ゆえに日本を代表する多くの政治家が交友関係にあり、盆栽文化の発展の礎を築いた多くの人物である。天皇陛下の盆栽も管理し、日本の盆栽を世界に知らしめる礎にもなった。

その村田先生は特にアメリカに盆栽を広める為、多くの努力をされてきた。そして日本の吉村香風園の子息である吉村氏をアメリカに盆栽の指導者として紹介し、後にこの方はアメリカへ移住し、盆栽学校を作り、多くの日系人や白人のアメリカ人に本格的盆栽を指導し、アメリカで盆栽文化発展の為に一生涯、一筋に盆栽を広めた。

その生徒の中に日系人のジョン仲氏がおり、後にアメリカ盆栽の礎を作り、アメリカ盆栽の生みの親と慕われ、アメリカ全土に多くの盆栽人を育てられるのである。そしてこの方が世界の盆栽の礎を作り、活動し、世界に多くの盆栽指導者を送り出す原動力となったわけである。

私は1970年に村田久造氏に依頼されずね、氏と共にアメリカ西とアメリカ東に盆栽の指導を行ったきっかけとなれが私が最初にアメリカへ盆栽の講演をして廻った。ジョン仲氏をたずね、その後数回アメリカへジョン仲氏の招きで指導に行った。後に木村正彦氏をアメリカへお連れし、B.C.I 1989年アメリカで盆栽のデモンストレーション企画し、ジョン仲氏の通訳により、多くの愛好家より喝采を受け、現在のアメリカ盆栽の発展の礎になるわけである。

（2）日本盆栽が世界盆栽にどのような影響をもたらしたか、その本質的原因

これは村田久造、吉村香風園、そしてジョン仲氏の功績であり、B.C.Iの設立と発展が私であると思う。その原動力になり多くの日本の盆栽作家がアメリカで研修、講演をし、アメリカ盆栽の発展が世界盆栽の発展に繋がったものであると考える。特にジョン仲氏の活躍は大変素晴らしいものと思います。現在の世界の盆栽の発展はジョン仲氏をおいて語りえない。

そして日本盆栽の美と芸術性、日本文化の素晴らしさが多くの世界の人達の感動を呼んだものである。「日本盆栽美学 侘 寂 の美学は究極のテーマとなっておる。」

**（3）近年日本盆栽は世界盆栽界での発展について日本の盆栽が世界の盆栽界の主流であることは永遠に変わることはないであろう。しかし、中国盆景の発展こそ、世界の盆栽のさらなる発展の大きな原動力となり、盆景の発展はジョン仲氏のグローバル化がなされ、さらなる発展を遂げると思う。

中国盆景が世界の盆栽の未来に与える影響は絶大であり、中国盆景が世界の盆栽の未来を変えることに思う。中国盆景の歴史、思想、宗教、哲学、美学は理論的で合理性に豊み、世界の人達に理解しやすいと考えている。中国盆景の発展は日本盆栽の発展となり、世界盆栽の発展になると信じている。

① 中国での議題
㋐ 盆栽と道教との関係
㋑ 天人合一は人は何をなすべきか

③意境とは
④山水思想とは
⑤盆景の発生目的（始皇帝の盆栽と盆景）
⑥庭園の理念と盆景の理念の関係
⑦盆景は何故2000年以上存在したか
⑧清朝 乾隆帝は盆景に何ほどの関心を持って、何の為に、特に黄帝は盆景に何を求めて盆器を作らせたのかそしてこの時代の盆景の発展は…

素晴らしい盆景となる。

日本の盆栽は中国から学ぶものは沢山あるが、教えとなるものは何もない。日本の盆栽は中国盆景を日本の土壌で養育し、成就させたもので、本質的には何も変わっていない。道教、儒教、仏教（禅）が日本盆栽の根本理念であり、中国盆景理念と相違することはない。中国神仙思想、不老長寿の概念は日本盆栽の根本思想、そして侘、寂、の美学に共通するものを感じる。

水石について

これは私が先人の言葉を引用し水石美についてまとめたものである。参考にしていただければ幸い。私の水石の概念となる。

水石は山水・景情を想い自然美を楽しむものである。

そして山水とは単なる山や川でなく、俗界に対峙する精神的象徴であり、哲学・宗教・歴史・文学・美術等の文化的価値と不可分に結びついた存在であり、水石は本質的には、山水画や仮山（山水庭園・枯山水）同じ目的で観照されるもの、山水を表面化したものが山水画であり、立体化したものが仮山であり、盆上におさめてしまった物が盆石（水石）である。水石は立体山水画、無声の山水詩と言われている。この場合の山水は単なる自然の風景とは異なり、それは非常に広い概念であり、自然の風景（真山水）のみならず胸中の山水、自然の風景山水、宗教的山水（神仙、道教、蓬莱山や仏教の須弥山など）等を含む。

山水とは中国で発達した極めて精神性の高い概念である。

すでに孔子は「知者は水を楽しみ（論語）仁者は山を楽しむ」と言っている。

六朝時代の宗炳（375～443）は「山水に致りて

走进宝岛台湾的盆栽艺术
Entering Bonsai Art in Treasure Island, Taiwan

2013 唐苑的世界盆景对话
DIALOGUE TO THE WORLD PENJING

发言人：梁悦美 图文整理：CP　Speaker: Amy Liang　Reorganizer: CP

第18届华风展盛况
The 18th HWA-FONG Exhibition event

华风展展场一角
A corner of HWA-FONG Exhibition

今天，很高兴可以和各位专家在这里讨论国际盆景。首先，请允许我介绍中国宝岛台湾的盆栽。

在清朝时，台南府城、鹿港、艋舺一带就有文人雅士赏玩盆栽。台南开园寺、鹿港妈祖庙、台北龙山寺保留至今的盆栽及盆钵，都有200年以上的历史。

宝岛台湾得天独厚，气候温和，湿度适中，四季如春，地处热带及亚热带，高山地区又有温带、寒带特质，所以无论热带、亚热带、温带甚至寒带野生林相，均一应齐全。天生地养也因气候好，发育快、生长期长、整姿容易、成形迅速。

几十年来，台湾不断地推广绿化运动。特于1984年5月20日，推出公共电视节目（社教节目）"中华盆栽艺术"，播出长达2年之久。在新闻局力邀及盆栽界公推之下本人担任主讲人，全年在电视上讲授及创作盆景。由于观众反应热烈，爱好者激增，观众累计达30万人以上。

一、中华盆栽艺术台湾总会

中华盆栽艺术台湾总会的会长每2年一届，届满换会长。从第一任总会长蒋金村医师。第二任总会长梁悦美到现在，至今已经第18届了。现全台湾有28个盆栽地方协会，每个地方协会每年都举办一次盆景展，经评审选出最好的10名盆栽入选华风展，加上顾问及企业收藏家邀请展出的盆栽和贵重盆栽等，共有300多盆盆栽在华风展上展出。华风展每年展出一次，每年约有10余个国家代表来参加，同时在展会上举办创作示范及讲演，每年讲师约4人，全部是由中国台湾推选出来的盆栽大师担任。数年来本人均被安排在第一位示范创作及讲解，为使各国盆栽友人更易理解，所以本人用国语、台语、日语、英文讲解。作为中华盆栽艺术台湾总会第二届总会长及15年来唯一的中华盆栽艺术台湾总会国际主席，我每年将华风展的邀请函及300本华风展的书，用中华盆栽艺术台湾总会的名义广寄赠送到世界各国的盆栽组织。

二、带领台湾盆栽代表团走向世界

目前国际间最具影响力的世界性盆栽雅石组织团体共有4个，ABFF亚太盆栽友谊联盟、WBFF世界盆景友好联盟、BCI国际盆景俱乐部、ASPAC亚太盆栽水石大会。

我15年来积极带领台湾盆栽代表团参加国际大型展及各个国家盆栽大展。组团参展的有：2004马来西亚主办的ASPAC国际展；2005菲律宾ASPAC国际展；2006中国北京ASPAC国际展；2007印尼ASPAC国际展；2011日本ASPAC国际展；2011泰国ABFF国际展；2013菲律宾ABFF国际展；2013中国扬州BCI国际展；2013中国金坛WBFF国际展；2013中国古镇中国鼎国家展等。

The Annual Forum 国际年度论坛

梁悦美，中国盆景艺术大师，国际首席盆栽大师，ABFF亚太盆栽联盟前会长，中华盆栽艺术台湾总会前会长，中国盆景艺术家协会名誉会长。

Being honored with the titles of "Chinese Penjing Art Master" and "International Chief Bonsai Master", Amy Liang is also the former President of ABFF (Asia-Pacific Bonsai Friendship Federation), the former President of Chinese Taiwan Bonsai Art Association and Honorary Chairwoman of China Penjing Artists Association.

每次台湾代表团我都亲自率团参加，人数都超过25人甚至到120人以上，团队表现优异，载歌载舞，非常有亲和力，所到的地方皆赞赏有加。此外各主办国如果只有邀请我个人或台湾团队无法成团，我都亲自一人代表中华盆栽艺术台湾总会前往参加。

三、中华盆栽艺术台湾总会举办的国际展

在台湾举办过ABFF国际展（邀请到16个国家、600余位国外盆栽专家），两次BCI国际大展（邀请到14个国家、数百余位国外嘉宾），第五届、第十届两次ASPAC国际大展（邀请42个国家，约1700位外宾）。每次国际大展都是盛况空前，引发世界盆栽热潮。在台湾盆栽人的共同努力下，一次次地克服一切困难，努力冲刺，经过千辛万苦，终于使台湾的盆栽享誉世界，在国际上倍受尊重与赞许。

另外，我的台北盆栽——紫园，接待国内外盆栽雅石贵宾，每年平均约3000人。只要到紫园的中外来宾，我都热忱款待，他们都感到"宾至如归"、"不虚此行"、"心满意足"。我一心一意就是要使全世界的盆景人，在台湾有一个温暖的盆景家。

台湾盆栽，是多元化的，思考较开放，视野较开阔，经过多年来的努力，已渐渐走出自己的风格。它融入中国园林哲学的思想、日本的精密纤细造型与植物管理学及西方现代美学的观念，又加上近年来以现代科学、植物生理的维护及管理方法，台湾已经有属于台湾本土独具一格的"台湾盆栽"了。

盆景是超越国界的艺术，它营造出一个绿色的和平世界。我和盆景结缘已经将近40年，盆景治愈了患有严重忧郁症的我，因此，我对盆景的付出是无怨无悔、矢志不渝的，终身为盆栽奉献。我前后在美国西雅图太平洋大学及南区大学，国立台湾师范大学、中国文化大学四所大学担任园艺盆景教授，出版的盆栽书有10本，在21个国家做过演讲，国内外剪彩140次以上。由1990年开始23年来我都自费来中国大陆讲授盆栽，中国园艺盆景培训班从第一届到第十三届我从未间断，为到大陆授课多次向美国我授课的西雅图太平洋大学请假一星期，据统计我在中国大陆有6400位学生。23年来中国盆景友人对我的温馨及关怀、浓情厚谊，深深感激、铭感肺腑。我真诚地期望盆栽不分国界、跨越派别，成为世界的共同语言与文化。40年的盆景生涯，在一个女人的一生中是何其漫长！？其中的酸甜苦辣，点滴在心头。愿以40年我的青春，换取我最心爱的盆景永远灿烂的生命！

华风展企业家展区——梁悦美理事长特展
Entrepreneurs corner of HWA-FONG Exhibition, for Amy Liang

华风展展场一角
A corner of HWA-FONG Exhibition

2013 唐苑的世界盆景对话
DIALOGUE TO THE WORLD PENJING

Entering Bonsai Art
in Treasure Island, Taiwan

Speaker: Amy Liang Reorganizer: CP

Today, I am very glad to have this opportunity to discuss the world Penjing with the experts. At first, please allow me to introduce bonsai in treasure island, Taiwan.

Back in the Qing Dynasty, refined scholars in the areas of Tainan prefectural city, Lugang and Mengxia were interested in bonsai appreciation. The bonsai and pots reserved in Kaiyuan Temple of Tainan, Mazu Temple of Lugang and Longshan Temple in Taipei are with the history of more than 200 years.

Treasured island of Taiwan is richly endowed by nature with gentle climate and appropriate humidity. It's like spring all the year round. Since it locates in the tropical and subtropical zones, and the high mountain region is with the qualities of temperate and cold zones, there are complete wild forest forms of tropical, subtropical, temperate and even cold zones. Due to the favorable climate, the plants grow fast with long growing period and are easy for modeling and quick for shaping.

In the last a few decades, Taiwan has continuously promoted the greening campaign. And a public TV show (a social education program) named Chinese Bonsai Art has been broadcasted on May 20, 1984 and lasted for 2 years. I was invited by the News Bureau and elected by the bonsai circle to be the keynote speaker to teach and create Penjing on TV all the year round. Due to the enthusiastic responses from the audiences, the number of enthusiasts rapidly increases, and the audience number reached more than 300 thousand people.

I. Chinese Taiwan Bonsai Art Association

The president tenure of Chinese Taiwan Bonsai Art Association is of 2 years, and the president changes at expiration. From the first president Dr. Jiang Jincun, the second president Prof. Amy Liang, there have been 18 tenures till now. Currently

梁悦美教授在华风展上示范表演
Demonstration by prof Amy Liang on the HWA-FONG Exhibition

there are 28 local associations in Taiwan, and every local association will hold a Penjing exhibition every year. The elected best 10 bonsai will be selected for HWA-FONG Exhibition. With bonsai and expensive bonsai invited by the consultants and enterprise collectors, there are totally more than 300 bonsai being exhibited in HWA-FONG Exhibition. HWA-FONG Exhibition is held once a year. Representatives from about 10 countries will participate in and perform creation demonstration and speech in the exhibition. There are about 4 lecturers every year and all of them are professional masters from Taiwan. Over the past few years, I have been the first to demonstrate the creation. In order to help friends from each country to understand more easily, I deliver the speech in Chinese, Taiwanese, Japanese and

寿娘子 飘长135cm 梁悦美爱培
Premna serratifolia. Branch: 135cm. Trained: Amy Liang

"金玉满堂庆新年"八房柑 飘长90cm 梁悦美培
"Gold and jade fill the hall to celebrate the New Year". Bafang Kumquat. Branch: 90cm. Trained: Amy Liang

翠米茶 飘长90cm 梁悦美爱培
Eurya emarginate. Branch: 90cm. Trained: Amy Liang

杜鹃 高80cm 梁悦美爱培
Rhododendron simsii. Height: 80cm. Trained: Amy Liang

The Annual Forum 国际年度论坛

English. As the second-term Chief President of Chinese Taiwan Bonsai Art Association and only International President of Chinese Taiwan Bonsai Art Association in the past 15 years, every year, I send the invitation letters and 300 books of HWA-FONG Exhibition to bonsai organizations all over the world on behalf of Chinese Taiwan Bonsai Art Association.

II. Lead Taiwan Bonsai Delegation to the world

Currently, there are 4 world bonsai and suiseki organization groups with the most influence in the world, such as ABFF (Asia-Pacific Bonsai Friendship Federation), WBFF (World Bonsai Friendship Federation), BCI (Bonsai Clubs International) and ASPAC (Asia Pacific Bonsai and Suiseki Convention & Exhibition).

For the last 15 years, I have actively led the Taiwan Bonsai Delegation to participate in great international exhibitions and the grand exhibition of various countries. We have formed the group and participated in the exhibitions as: the ASPAC International Exhibition sponsored by Malaysia in 2004; the 2005 ASPAC International Exhibition held in Philippines; the 2006 ASPAC International Exhibition held in Beijing of China; the 2007 ASPAC International Exhibition held in Indonesia; the 2011 ASPAC International Exhibition held in Japan; the 2011 ABFF International Exhibition held in Thailand; the 2013 ABFF International Exhibition held in Philippines; the 2013 BCI International Exhibition held in Yangzhou of China; the 2013 WBFF International Exhibition held in Jintan of China; and the 2013 China Ding -- China National Penjing Exhibition held in Guzhen of China, etc.

紫园一景
A view of Purple garden

I have led Taiwan delegation in person every time, and the number of member all exceeds 25 people, even above 120 people. The team has performed excellently. They sing and dance with great appetency and are appraised highly in every country they have been to. If the host country only invited me or the Taiwan delegation cannot form a group, I would participate in the exhibition on behalf of Chinese Taiwan Bonsai Art Association all by myself.

III. International Exhibition have been held by Chinese Taiwan Bonsai Art Association

Taiwan has held ABFF International Exhibition (more than 600 foreign bonsai experts from 16 countries have been invited), two BCI International Exhibitions (hundreds of foreign guests from 14 countries have been invited), and the 5th and the 10th ASPAC International Exhibition (about 1700 foreign guests from 42 countries have been invited). And every time, the splendor of the occasion surpassed anything heretofore seen and triggered the upsurge of world bonsai. Under the joint effort of Taiwan bonsai artificers, we have

1976年，美国200周年国家庆典之际梁悦美教授在美国华盛顿国家公园教学 In 1976, American 200 anniversary of National Day celebration, Prof. Amy Liang is teaching in the United States Washington D.C. National Park

梁悦美教授在教授盆栽
Prof. Amy Liang is teaching in bonsai

Entering Bonsai Art
in Treasure Island, Taiwan

紫园后院的盆景水墙与花卉一隅
The Penjing water wall and a corner of flower in purple garden's backyard

overwhelmed all the difficulties and strived to exert again and again. After innumerable trials and hardships, we have finally made Taiwan bonsai renowned by the world and have won it respect and appreciate of the international community!

Additionally, my Purple Garden in Taipei has received about 3000 bonsai and suiseki guests from home and abroad averagely every year. When the Chinese and foreign guests came, I sincerely treat, they all feel "at home", "worthwhile", "satisfied". I have devoted my heart and soul to afford a warm Penjing house for all the Penjing artificers over the world.

Taiwan bonsai is diversified, and the thought and view are relatively open. After the efforts of many years, it has gradually formed its own style. It has blended in the thought of Chinese garden philosophy, the precision and slender modeling and plant management of Japan, as well as the view of western modern aesthetics. In addition with the modern science, maintenance and management method of plant physiology, Taiwan has owned the unique and native "Taiwan Bonsai".

peaceful world. I have become attached to Penjing for almost 40 years. My serious melancholia has been cured by Penjing. Thus, my contribution for Penjing is without regrets and unchangeable. I have worked as gardening bonsai professor in U.S. Seattle Pacific University and South Seattle Community College, in National Taiwan Normal University and Chinese Culture University. I have published 10 books, gave lectures in 21 countries, cut the ribbon at home and abroad more than 140 times. In addition, since 1990 I have given lectures about bonsai in mainland of China every year in the past 23 years, every time I paid by myself. From the first to the 13th China garden bonsai workshops, I never stopped teaching. For it, I had to ask United States Seattle Pacific University for a week off many times. According to statistics, I have 6400 students in mainland China. In 23 years, the mainland Penjing friends show the most friendly and the warmest solicitude for me, I am deeply grateful, and I'll remember the deep friendship forever. I sincerely hope that bonsai will become a common language of the world beyond borders and factions. For the condensed art of nature and our common enthusiasm, let us link the bonsai culture and friendship all over the world through bonsai! And look forward to the future, people all over the world proud of "I have bonsai". In the 40 years of Penjing, how long it is in a woman's life. I feel every little of the past happened joys and sorrows of life. I'm willing to exchange 40 years of youth for innortal brilliant life of my favorite Penjing.

紫园夕景
Purple garden's evening scene

The Annual Forum 国际年度论坛

2013 唐苑的世界盆景对话
DIALOGUE TO THE WORLD PENJING

盆景大小之我见
My Opinion on the Size of Penjing

发言人：杨贵生　图文整理：CP　Speaker: Yang Guisheng　Reorganizer: CP

大家好，我是来自中国江苏苏州的盆景爱好者。今天我想跟大家探讨的议题是为什么中国盆景发展的趋势是越来越大。

盆景在中国有着悠久的历史，从传统的角度来说，盆景的高度都在30～90cm之间，但是随着时代和经济的发展，现在盆景的发展出现了越来越大的趋势。这是为什么呢？这是因为中国的经济在慢慢地发展，大家的审美观也在发生变化，中国私人园林的规模越来越大，庭院越来越大，如果在一个很大的庭院里面布置传统高度、相对很小的盆景，观赏起来视觉效果就会有很大落差，比如唐苑这么庞大的园林，摆放传统高度的盆景就不般配了。所以这次中国盆景艺术家协会在广东中山举办的盆景展览必然会出现参展盆景规格很大的这种现象。

我觉得盆景是大还是小，这没有任何对错之分，大有大的好，小有小的好，这是每个人根据自己的经济条件和个人喜好来决定的。我认为，大盆景也有它的缺点，就是大众化很困难，因为大盆景从经济价值、养护等客观条件来说都是要求比较高的。

那么如何解决大盆景和小盆景之间的矛盾呢？首先，我觉得各个盆景专家也好，各盆景协会也好，都需要给大众做出正确的引导。第二，今后各级盆景协会举办展览的时候应该注意要多元化。将大盆景和小盆景应该分开展览，否则会有大盆景"欺负"小盆景的嫌疑。第三，在以大盆景为主流收藏的现实条件下，也要大力提倡小盆景，毕竟小盆景才是大众化的。

我个人收藏的计划是大型盆景占40%，中型盆景占50%，小型盆景占10%。

杨贵生，中国盆景艺术家协会常务副会长。
Yang Guisheng, the Executive Vice-President of China Penjing Artists Association.

Hello, everyone, I'm a Penjing lover from Suzhou, Jiangsu, China. Today, I want to discuss with you the topic "Why the size of China Penjing becomes larger and larger".

Penjing has a long history in China; from the traditional view, Penjing is 90cm~30cm high. However, with the development of time and economy, the current Penjing becomes larger and larger. Why? The reason is that: With the economic development of China, our aesthetic standard also has changed; in addition, both the Chinese private gardens and courtyards become larger and larger. If one lower and smaller Penjing is arranged in a larger courtyard, it will bring a larger visual effect gap. Take Tangyuan, a huge garden, as an example, if a lower and smaller Penjing is arranged, it will bring a mis-matched effect. Accordingly, it is inevitable that China Penjing Artists Association will show larger Penjing in Penjing exhibition in Zhongshan, Guangdong.

I think there is no right or wrong answer for Penjing size, both of them have their own advantages, which will be determined by our economic conditions and personal preference. I think that the large Penjing will also have its own disadvantages: As the large Penjing has the stricter requirement on economic values, care and other objective conditions, its popularization will be very difficult.

So how to solve the conflict between the large Penjing and the small Penjing? Firstly, I think both the Penjing expert and Penjing association shall give proper guidance for all the people. Secondly, higher attention shall be paid to the diversification of exhibitions held by all Penjing associations. The large Penjing and the small Penjing shall be exhibited in a separate way; or else, the large Penjing will "withhold" the small Penjing visually. Thirdly, in the condition of taking the large Penjing as the essential collection, the small Penjing shall also be vigorously advocated, for it is popular.

According my personal collection plan is that 40% of my collection are large Penjing, 50% medium Penjing and 10% small Penjing.

盆景的精神

发言人：【法国】克里斯蒂安·弗内罗；米歇尔·卡尔比昂　图文整理：CP
Speaker: [France] Christian Fournereau & Michèle Corbihan　Reorganizer: CP

第一个运抵法国的盆景
First Penjing that arrives at France

海棠 *Malus sieboldi* 高 80cm
吉尔伯特·勒布里德藏品
Crap-apple. Height: 80cm.
Collector: Gilbert Labrid

圆叶樱桃 *Prunus mahaleb*
高 82cm 蒂埃里·丰藏品
Cherry. Height: 82cm.
Collector: Thierry Font

软枣猕猴桃 *Actinidia arguta*
高 52cm 乔尔·普泽藏品
Height: 52cm.
Collector: Joel Pouzet

榆树 *Ulmus.* 高 57cm 皮埃尔·埃罗藏品
Elm. Height: 57cm. Collector: Pierre Hérault

Jean-Claude Febb
让·克劳德·费伯藏品

　　亲爱的盆景爱好者们，对我们来说能参加首届 2013 "唐苑的世界盆景对话"国际年度论坛是一个极大的荣幸。我们来自法国，并担任法国《气韵盆栽》杂志的主编和出版人。今年这本杂志已经成功出版了十周年，并且它也是法国阅读量最多的杂志。

　　《气韵盆栽》是指或多或少的"盆景的精神"。我们选择这个作为我们杂志的名字，是因为我们想要呈现"盆景"的艺术，并探索所有与盆景相关的艺术，例如：陶器、插花、盆景树木用盆的收藏等。

　　《气韵盆栽》确实提高了盆景树木所能带来的令人惊叹的美，并向读者们提供了必要的解读，让他们学会如何展示他们作品最好的一面。

　　最早出现在法国的盆景是在 1889 年巴黎世界博览会的盆景，它们都来自日本。

　　法国的盆景艺术缓慢发展于 19 世纪 60 年代，在 19 世纪 80 年代末开始变得越来越重要。

　　在法国和欧洲日本的盆景众所周知，但是中国盆景影响力却仍然不足，因为没有太多能获取到中国盆景信息的渠道。我们不时地在《气韵盆栽》杂志上发表几篇关于中国盆景的文章，我们知道还是有很多对它们感兴趣的读者。

　　威利·本茨（Willy Benz），一个伟大的德国作家和中国艺术业余爱好者，经常撰写关于盆景以及自己对宝成博物馆、盆景艺术的意义以及盆景本身等方面的文章。

　　《气韵盆栽》团队第一次看到中国盆景是在位于法国西南部的一个小镇——鲁瓦扬的博物馆。对我们来说非常有趣的事是发现这棵大尺寸的盆景树我们对它大小的印象非常深刻，它比我们所知道的日本盆景大出很多。

　　法国盆景的灵感绝大多数来自日本盆栽艺术，只存在稍许不同。法国最受欢迎的树种是位于法国南部的松树、枫树和橄榄树。法国人所喜爱的盆栽树也多来自于当地树种，例如：橡树、常春藤、榆树……，他们也对"能开花的树"（例如：杜鹃花……）有极大的兴趣。

　　法国的盆景或盆栽由法国盆栽协会（FFB）进行组织，下面又有 120 个协会和大约 3000 位成员。在法国，每年大约会举行 10 场重要的展会，来自世界各地的艺术家以及来自日本的大师们都会参与该展会并来展示他们的盆景艺术（与中国的展会形式相同）。当地的俱乐部每年也会举行许多其他较小的展会。

The Annual Forum 国际年度论坛

克里斯蒂安·弗内罗，法国《气韵盆栽》杂志出版人；米歇尔·卡尔比昂，《气韵盆栽》杂志主编

Christian Fournereau, the publisher of Esprit Bonsa; Michèle Corbihan, the redactor-in-chief of Esprit Bonsai

    FFB每年都会组织全国性展会，每次都会在法国不同的地方的举行，旨在让其遍布法国的每个地区。法国盆景艺术家也参与到国际活动中，例如：在国际上最享有盛名的比利时的"Noelander's Trophy 诺朗德斯杯"。

    FFB正在努力工作以期对他们的成员以及盆景专业人士进行培训。这个想法是为了增加盆景知识以及提高盆景制作的技术……同时我们也希望能增加盆景爱好者的人数。

    现在，在法国进行盆景实践主要是受日本盆景制作方式的影响。我们希望在将来，我们能更多地采用中国方式来对盆景进行加工制作并带来一个能让所有人都受益的新的视野。

    我们的《气韵盆栽》杂志试图成为世界盆景和法国和世界各地盆景进化史的"忠诚"见证者。所以，我们会非常乐意在我们接下来的刊物中发布第一份关于中国盆景的报告并报道有关在扬州举行的国际盆栽俱乐部50周年庆典的相关活动。

    同时，我们也感到非常荣幸可以参与此次的中国盆景国家大展和2013"唐苑的世界盆景对话"国际年度论坛，让我们有机会能欣赏在盆景史上最高水平的展会并发现来自中国各地的不同类型的盆景。

我们将非常荣幸能在我们的杂志上报告这次精彩的活动，很高兴能借此机会来提高中国盆景在法国的知名度并增进我们两国之间的友谊。

    谢谢各位。

Dear Friends of Penjing, It is a great honor for us to participate at this first Annual Forum of "the 2013 Dialogue to the World Penjing".

We are coming from France and are the Chief Editor and the Publisher from the French magazine "Esprit Bonsai". This magazine is celebrating this year his 10th anniversary and is the most read magazine in France.

"Esprit Bonsai" means more or less "Spirit of Bonsai". We choose this title for our magazine, because we wanted to present the art of "potted trees" and explore all the arts related to the bonsai such as pottery, flower arrangement, potted collection of plants to name only those.

"Esprit Bonsai" really enhances amazingly beautiful trees and provides the reader with the necessary explanations to make their own and learn how to present them at their best.

The first Penjing arrived in France in 1889 for the universal exhibition in Paris. They came from Japan. In our country, the art of Penjing began to develop slowly in the 1960s and in a more important way at the end of the 1980s.

In France, and in Europe, Japanese Bonsai are well-known, but Chinese Penjing are still underestimated, because we do not have lots of information about them. From time to time we published in our magazine few articles about Chinese trees and we know that lots of our readers are interested in them.

Willy Benz, a great German writer and amateur of Chinese arts, wrote very often about Penjing and shared his knowledge about the Baocheng museum, the meaning of the art of Penjing, and the Penjing themselves.

The first time the Esprit Bonsai team saw Chinese trees was in the museum of Royan, a small town in the south west of France. It was very interesting for us to discover these big trees, and we were very impressed by the size of them, which were much taller than the Japanese one we knew…

Potted trees in France are very closely inspired from Japanese Penjing art, with a few differences, according to the essence of the trees. The most popular trees that are formed in France are pines and maples, and also "olive-trees", in the south of France. French lovers of potted trees also form local trees like Oaks, charms, or elms… and have a great interest for "flower trees" like azaleas…

Penjing, or bonsai, in France, is organized around the French Federation of Bonsai, "la Fédération Française de Bonsai (FFB), with more than 120 associations and approximately 3.000 members. There are in France around 10 important exhibitions every year, with artists from international matter and Japanese masters coming to organize shows about their art, like it happens in China too. And there are lots of smaller exhibitions organized by the local clubs all around the year.

The FFB organizes every year her national convention, each time in a different place, trying to be presented in each French region. French bonsai artists are also taking part to international events, like the "Noelander's Trophy" in Belgium, which is probably internationally the best known of them.

The FFB works very hard to develop the training of their members as well as the professionals of Penjing. The idea is to increase the knowledge, the quality of the practice and of the trees… and we hope, the numbers of the Penjing lovers.

Nowadays, the practice of Penjing in France is mostly inspired by the Japanese way of work. We hope that in the future, the Chinese way of dealing with trees will increase and will bring a new vision that will benefit to all.

Our magazine "Esprit Bonsai" tries to be the "faithful" witness of the world of Penjing and its evolution in France and all around the world. So, we would be very happy to publish a first report about bonsai in China in our next issue and to present the exhibition around the 50th anniversary of the Bonsai Club International hold in Yangzhou.

And we are very happy to be with you this evening, very happy to have been chosen to visit one of the most important exhibitions of Penjing in China and to admire this exhibition with the highest level in Penjing history and discover Penjing of various kinds from all over China.

We will be glad to report about these wonderful events in our magazine and we are very happy to get the opportunity to contribute to increase the popularity of Chinese Penjing in France and to the rapprochement and friendship between our two countries.

We will be glad to report about these wonderful events in our magazine and we are very happy to get the opportunity to contribute to increase the popularity of Chinese Penjing in France and to the rapprochement and friendship between our two countries.

Thank you for your attention.

天使的传递
Angel's Transmission

发言人：【立陶宛】凯斯图蒂斯·帕特考斯卡斯 图文整理：CP
Speaker: [Lithuania] Kestutis Ptakauskas Reorganizer: CP

凯斯图蒂斯·帕特考斯卡斯 欧洲盆栽协会成员国立陶宛会长
Kestutis Ptakauskas President of Lithuania of EBA Member Country

欧洲盆栽协会成员国立陶宛会长凯斯图蒂斯·帕特考斯卡斯（右）赠予盆景天使给中国盆景艺术家协会会长苏放（左）
Kestutis Ptakauskas, President of Lithuania of EBA Member Country (right) send an angel with Penjing to Su Fang, President of China Penjing Artists Association (left)

　　致尊敬的苏放先生，国际论坛的组织者及全体与会人员，首先，我要感谢主办方和苏放先生能给我这个难得的机会让我能为这次活动尽一份力。

　　这是我第一次来到中国，这个有着几千年盆景艺术历史的国家。就盆景艺术这方面而言，立陶宛看起来就像一个刚出生的婴儿，我们只有一位盆景大师和几位盆景爱好者。然而，我们正在组织一次高级别的国际盆景展会，并会邀请来自世界各地的顶级盆景大师们。

　　现在，我想和你们分享我自己的故事，这也正是立陶宛盆景历史的开始。正如您所看到的，我手上正拿着一个天使，她手里捧着一盆小盆景。她看上去就像守护着我的天使。今天我愿意把她送给苏放先生，以此见证我们两国的盆景友谊。

　　每天早上我都会在一个名叫"朝露花园"的园子里冥想，"朝露花园"是我亲手打造的盆景园。然而，我也为这个花园的建造付出了代价和艰辛的努力。

　　我出生在一个被驱逐出境的西伯利亚家庭。我的家人设法返回立陶宛，因此我们不得不重新开始一切并克服许多障碍。但是事情却变得更糟，作为一名苏联军队的士兵，我被迫去参加阿富汗战争。战争在我的身体和灵魂上留下了许多烙印。它给我留下了一个血流不止且不能愈合的伤疤……我不得不找一些能让我的生活变得有意义的事情，并让它帮助我在生活中找到平衡。起初时，我尝试每天都让自己感到疲劳，这样的话我就不会去思考其他的事情。我只能睡几个小时，剩下的时间我让自己在工作和学习中度过。然而，不幸却从未结束。我设法从五次劫难中幸存下来。于是我开始明白，有人一直在保护我同时鼓励我继续寻找能帮助我度过各种生活苦难的方法。

　　1989年，我从一位朋友那里购买了我的第一盆盆景。这就是我对这个小小的神奇世界越来越感兴趣的开始。我在森林里寻找它们，并在多年的精心养护后将它们变成这些奇迹……在我的收藏中，有些盆景在国际展会上获得奖项。我很高兴的是，其中几盆盆景被收入《世界金奖盆景集》一书中，该专辑于2006年在中国出版。

　　我的花园里有一个茶室，在那里我的客人不仅可以品茶，还能和我一起思考流逝的光阴以及人生存在的意义。似乎一杯茶可以包含的东西远远不止这个世界……当我的花园建成以后，各行各业的人们都开始前来参观——他们都是善良的人们，有着平静的灵魂，对传统的东方文化和朴素的民风感兴趣的人们，在此寻求志同道合的朋友。

　　有人会问，我们这样的小国家应该怎样设法去找到这么多对盆景感兴趣的人呢？2003年我建立了盆景工作室在我们国

The Annual Forum 国际年度论坛

凯斯图蒂斯·帕特考斯卡斯每天清晨在自己创造的"朝露花园"打坐
Kestutis Ptakauskas meditates in his *Morning Dew* garden every morning

凯斯图蒂斯·帕特考斯卡斯的"朝露花园"及茶室
Kestutis Ptakauskas's *Morning Dew* garden and tea house

家传播盆景艺术。每年有25名学生由来自全球专业的盆景大师授课。但我们面临的最大问题是我们没有盆景大师,不得不花昂贵的费用从其他国家邀请艺术大师为学生们授课。这就是为什么盆景艺术在立陶宛发展的很慢的原因。我个人认为,有许多专业盆景大师的国家可以来帮助像我们这样缺乏师资力量,因为我们的人民渴望学习其他国家的盆景艺术。在过去的十年间,我在立陶宛组织了八个盆景水石展会。这些展会在国际上均得到了认可。

而且我很荣幸地告诉大家,立陶宛将筹备2015年欧洲盆栽和水石大会以及2016年国际盆栽俱乐部大会,届时将会邀请来自35个国家的盆景大师。我为能邀请你们来访问我的国家而感到非常的快乐,让我们一起来分享对盆景和东方文化的热爱。

日本外交事务代表:金安艺卓(右一)和他的妻子,金安妙子(右二)
Eizo Kaneyasu (first on the right) and his wife, Taeko Kaneyasu (second on the right)

2004年第一届国际盆栽水石展在立陶宛阿利图斯举办
1st International Bonsai and Suiseki Exhibition in Alytus 2004, Lithuania

2005年第二届国际盆栽水石展在立陶宛阿利图斯举办
2nd International Bonsai and Suiseki exhibition in Alytus 2005, Lithuania

第三届国际盆栽水石展在立陶宛阿利图斯举办
3rd International Bonsai & Suiseki Exhibition, Alytus 2011, Lithuania

第六届国际盆栽和水石展上的盆景展品和水石展品
Exhibits of Bonsai and Suiseki on the 6th International Bonsai & Suiseki Exhibition

第八届国际盆栽水石展在立陶宛阿利图斯举办
8th International Bonsai & Suiseki Exhibition, Alytus Lithuania

Angel's Transmission

Speaker: [Lithuania] Kestutis Ptakauskas Reorganizer: CP

Dear honorable President Su Fang, organizers and participants of this international forum,

Firstly I would like to thank the organizers and President Su Fang for this amazing opportunity to be a part of this event. This is my first time in China, the country that has thousands of years history of Penjing art. Lithuania in this context looks like a newborn baby – we have only one Penjing master and several Penjing amateurs. However, we are organizing high-level international Penjing exhibitions and inviting world-class masters from all around the world.

Now, I want to share my story with you, which is the start of Penjing history in Lithuania. As you can see, I am holding an angel with a small Penjing. Perhaps it might symbolize my guardian angel. I'd like to send the angel to Mr. Su Fang to witness our two countries' Penjing friendship.

Every morning I meditate in *Morning Dew* garden - my own creation. However, I paid a price and arduous efforts for the construction of the garden.

I was born in Siberia, into the family of deportees. My family managed to return to Lithuania and had to start everything anew and overcome many obstacles. Things became even worse, when as a soldier for the Soviet army, I was forced to go to war in Afghanistan. War left its marks in my body and soul. It left a bleeding scar... I had to find something that would give me meaning and help me find a balance in life. At first I tried to tire myself so that I would not think of anything - I slept just several hours, and the rest of time I worked and studied. However, misfortunes never ended. I managed to survive five situations that could have led to death. And then I started to understand that someone is protecting me and at the same time encouraging me to continue my search for something that would help me withstand the hardships in life.

In 1989 I bought my first Penjing from a friend. That's the beginning of how my acquaintance with this tiny miracle of the world. I searched for them in the forests and turned them into the miracles after many years of careful maintenance. In my collection I have several Penjing that got awards in many international exhibitions. I am pleased that some of them have also been included into the photo album "Gold Awarded Penjing of the World" which was released in China in 2006.

My garden also features a teahouse, where my guests can not only have a cup of tea, but also think with me about the passing time, and the value of the moment. It seems that one cup of tea can contain more than the whole world... When my garden was created, various people started visiting it - those with good heart and calm soul, those interested in Eastern culture, traditions and customs, and those searching for their path, harmony in this world.

Some people perhaps ask how our little country manages to find so many people interested in Penjing. In 2003 I established a Penjing studio to spread Penjing art in our country. Each year there have been 25 students who have been taught by professional Penjing masters all over the world. The most important problem is that we don't have many masters and have to invite masters from other countries, which cost lots of money, and students' expense is too much for my studio. This is why Penjing art is spread slowly in Lithuania. Personally, I hope those countries that have lots of professional masters would help those countries like Lithuania that lack of teachers, but our people are eager to learn the art of Penjing. I organized eight Penjing and suiseki exhibitions in Lithuania in the past ten years. These exhibitions received recognition internationally

Also Lithuania has the honor to organize European bonsai and suiseki congress in 2015 and Bonsai Club International Congress in 2016 with Penjing masters from 35 countries. It is an enormous pleasure for me to invite you to visit our country, which shares the love for Penjing and Eastern culture.

The scenery of field event on the 8th International Bonsai & Suiseki Exhibition

印度盆景发展史
Penjing History in INDIA

发言人：【印度】苏杰沙 图文整理：CP Speaker: [India] Sujay Shah Reorganizer: CP

苏杰沙 印度盆景大师
Sujay Shah Indian Bonsai Master

20世纪60年代，阿格尼霍特利先生是第一个获得"帕德玛"奖的印度人（"帕德玛"奖是指颁给公民以感谢他/她对各知识领域的贡献的最高奖项之一）。这是第一次将一个全国认可的奖项颁发给一位印度的盆景艺术家——他是第一个将盆景艺术引进印度的人，也是这个领域的先驱者。在印度前总理英迪拉·甘地的指导下，制作了一部有关阿格尼霍特利先生的纪录片。

印度孟买的苏鲁普盆栽村
INDIA Ssurup Bonsai Village-Mumbai

印度由28个州和几个较小的联邦属地组成。每个州大约有5~7个不错的盆景俱乐部。有些是众所周知的，有些是刚成立不久的，但很快就会在国家舞台上显露头角。由于印度是一个如此广阔和多元化的国家，因此可以非常容易地将其划分为四个区域——北部、南部、东部和西部。

北部的盆景艺术始于1970年。最早的协会是由莉拉·达汉德太太在新德里成立的印度盆栽协会。在南部，维瑞克莎盆栽研究协会的苏杰拉、苏耶姐·柯棣华和娆拉哈是盆景艺术最显著的支持者。拉文德兰先生的纳盖科伊尔会所同样也拥有大量的盆景收藏品。东部几个杰出的盆景团体包括加尔各答盆栽协会、乔蒂迈俱乐部以及一些独立个人对盆景艺术的支持，他们是乔蒂迈俱乐部的曼珠·潘山瑞、乌纱·莫迪、冉敏·古普塔。已故的施里德先生和现在的帕哈·施里德夫人也是著名的盆景领袖。西部的印度孟买盆栽协会是由一位名为玛丽蔡斯的美国人成立的，并获得一群印度女性的青睐。（其中以苏尼塔·瓦斯瓦尼和拉特纳·萨丹为代表）。

乔蒂和尼坤·帕利于1979年成立了印度日本协会盆栽研究小组，这对盆栽艺术在西方的发展而言非常有帮助。它支持盆栽研究小组各项事务，并参与到将许多盆栽大师引入印度的展会的组织中。在20世纪80年代和90年代，这个小组是传播盆栽艺术的领军者，值得赞扬的是他们把印度盆景推向国际舞台。

乔蒂和尼坤·帕利的盆栽研究小组出版的《日清盆栽》杂志以及希恩·帕拉萨出版的《园艺世界》是非常值得关注的盆景出版物，其囊括了非常有价值的有关印度盆景艺术的最新动向。然而，就盆景刊物的出版和引进而言，媒体宣传渠道有提升的空间。

印度友谊盆栽协会由苏杰沙和乌瓦什·萨克在20世纪90年代成立于孟买。之后在2000年，苏杰沙和他的妻子苏鲁普一起开始创建苏鲁普盆栽园的培训小组，对盆景进行深入探讨。他们在鲁迪·纳乔安——印尼著名大师的帮助下创造了一个专业的盆景村。同时，还邀请了来自全球不同国家（例如：印尼、中国、泰国和波多黎各）的大师们。学生在这里可以学习的远远不止技术。他们到这里来可以体会到艺术背后的哲学理念、探索灵魂的真谛和盆景美学。苏鲁普盆栽花园是当代印

2013 唐苑的世界盆景对话
DIALOGUE TO THE WORLD PENJING

苏鲁普盆栽村创立者：鲁迪·纳乔安（左），苏鲁普（中），苏杰沙（右）
Surup Team: Rudy Najoan (left), Ruppa Shah (middle) and Sujay Shah(right)

用心感受美丽的景色
Man's heart feels serene in Ssurup Bonsai garden

灿烂阳光的照耀下每天都有爱的传递
Where the glorious rays of the sun shines each day spreading love

度盆景艺术现状的完美缩影。我们是首批在21世纪积极实践这一古老艺术的新生代。

印度气候为热带气候，特有的盆景为气生根的榕树盆景。我们有极其少量的山采树种，盆景创作上我们强调三角形，植物解剖学和整体的流动韵味。我们知道我们的盆景应该反映当地的植物群、文化和哲学。所有的国家都有其特有的气候特征、文化特色、哲学思想和技法，这些都将体现在他们的盆景作品中。印度的盆景艺术非常年轻，与其他国家相比还处于盆景艺术发展的初期阶段。

在我们看来，中国盆景非常迷人并让人感到眼前一亮，天衣无缝地将大自然通过盆景的表现带进家里。"对树的修剪和走势定位"的技艺令人难以置信，因为这种技法表现的树相尤为自然，没有一点人工雕琢的痕迹。我们印度制作盆景的技法有点墨守陈规，我希望我们向中国学习丰富的制作技艺，不仅学习盆景艺术还包括其他各行各业。

我们对盆景艺术的发展有许多复杂的感情。盆景不仅能让人放松，还能在这个充满压力的世界中舒缓压力。艺术的重要性正在全球兴起，某些国家的发展速度比其他一些国家更快。我们相信，如果艺术得到适当支持，它会产生巨大的发展潜力和增长。今天的主要障碍是指哲学理念，例如：中国风水学和印度风水学。此外，在印度，艺术主要是由女性实践，这也是一个严重的限制。然而，由于得到国际社会的支持，特别是来自快速发展的发展中国家（例如：中国）的支持，印度盆景艺术的前途是光明的。年轻的一代（男女双方）都需要接受盆景培训和教育。国际专家、教师和大师应该协助培训和发展，而且今后盆景的发展需要融入来自不同国家的新鲜血液和新技术。

感受多彩的生活
Touch of color and life

你可以感受到的温暖
Warmth you can experience

九重葛属植物
Bouganvilleas

苏铁科植物园
Cycad garden

Sujay Shah Indian Bonsai MasterIn the 1960s, Mr. V.P Agnihotri was the first Indian to be presented the 'Padmashree' award (one of the highest awards presented to a civilian for his or her contributions in various spheres of knowledge). This was the first time that a nationally recognized award was presented to a Penjing artist in India – he was the first to introduce the art of Penjing in India and was a pioneer in his field. A documentary was made about Mr. Agnihotri under the direction of the late Prime Minister of India, Indira Gandhi.

India consists of 28 states and a number of smaller union territories. Each state has about 5 to 7 good Penjing clubs. While some are well known, some are new and soon to emerge on the national stage. India is a country so vast and diverse that it is easiest to divide into four regions – the North, South, East and West.

In the North, the art of Penjing came alive in 1970. The very first association that was formed was the Indian Bonsai Association in New Delhi by Mrs. Leela Dhanda. In the South, Sushila and Sujata Dwarkanath, Latha Rao of the Vriksha Bonsai Study are prominent supporters of the art. Mr Ravindran's Nagercoil also has a vast collection of Penjing. In the East, prominent Penjing groups include the Calcutta Bonsai Association, Jyotirmai Club's Manju Pansari, Usha mody and Rashmi Gupta and several independent individuals have supported the art. Late shri Lala Shridhar and present Mrs. Prabha Shridhar are also the prominent figures in Penjing. In the West, the Indian Bonsai Society in Mumbai was formed by an American named Mary Chase and supported by a group of Indian women – the most prominent of whom were Suneeta Vaswani and

The Annual Forum 国际年度论坛

Penjing History in India

Speaker: [India] Sujay Shah Reorganizer: CP

Ratna Thadani.

Jyoti and Nikunj Parekh started the Bonsai Study Group of Indo Japanese Association in 1979. It was instrumental to the development of the art of Bonsai in the West. It supported all affairs of Bonsai Study Groups and was involved in the organization of exhibitions for which many Bonsai masters were brought to India. Thus in 1980s and the 1990s, this group was a leader in spreading the art of Bonsai and it deserves credit for putting India Penjing on the international stage.

The magazines – Nichin Bonsai, published by Bonsai Study Group of Jyoti and Nikunj Parekh, and Horticulture World by Sneh Prasar, are the magazines that are noteworthy and contain valuable information of the art of Penjing in India. However, there is scope for the introduction and publishing of many channels of media in the field today.

In the 1990s, the India Friendship Bonsai Society was formed by couple Ruppa and Sujay Shah and Urvashi Thacker in Mumbai. Later in the year 2000, Sujay Shah with his wife Ruppa Shah started Ssurup Bonsai Garden group for further serious training in Penjing. They created a Bonsai village professionally with the help of Rudy Najoan – a renowned master from Indonesia, and also invited different masters throughout the globe countries like Indonesia, China, Thailand and Puerto Rico. Students come here to learn much more than techniques. They come to understand the philosophy behind the art, the soul factor and the aesthetics. Ssurup Bonsai Garden provides the perfect snapshot of the current state of the art of Penjing in India. We are the first part of active generation to practice this ancient art in the 21st century.

India's climate is tropical and something that is unique to Penjing in our country is the aerial roots of the *Ficus microcarpa* Penjing. We have very negligible Yamadori and we emphasize more on taper, anatomy and flow. We have learned that our Penjing should reflect our flora, culture and philosophies. All countries have their own climate, culture and philosophy, techniques which come out in its art. The art of Penjing in India is very young. Compared with other countries, the development of Penjing art in India is still in the early stage.

In our opinion, China Penjing is fascinating and provides a 'wow feel factor' related to bringing nature seamlessly into homes. The technique of clip and grow method is fabulous as it gives a natural look to the tree and moves away from any sense of artificiality (manmade) aesthetics. We follow many of the same techniques in India but we have a lot to learn from China, not only about the art of Penjing but in many different walks of life.

We have mixed feelings of development of the art of Penjing. Penjing provides relaxation and is a stress buster in today's stressful world. The prominence of the art is growing up all over the world. Some countries' development is faster than that of others'. We believe that if the art is properly supported it has a tremendous potential for development and growth. The main hurdles today are philosophies such as Fengshui and Vastu Shastra. Moreover, in India, the art is practiced mainly by women, which is a severe limitation. However, with international support, especially from developing countries of fast developing like China, the future is bright. Young generations from both sexes need to have access to Penjing training and education. International experts, teachers and masters must assist with training and development. There must be a cross pollination of techniques and styles from different countries.

人工荷塘
Manmade water ponds

苏鲁普盆栽村每月研讨会
Monthly Meet of Ssurup

针对个人小组的制作表演
Demonstration to Private group

印度盆景欣赏
Appreciation of India Penjing

印度盆景欣赏
Appreciation of India Penjing

印度盆景欣赏
Appreciation of India Penjing

匈牙利国家盆景概述
Hungarian National Penjing Overview

发言人：【匈牙利】阿提拉·鲍曼 图文整理：CP Speaker: [Hungary] Attila Baumann Reorganizer: CP

匈牙利共和国位于欧洲的中部
The Republic of Hungary lies in Central Europe

匈牙利共和国国家概况
The country views of the Republic of Hungary

阿提拉·鲍曼 欧洲盆景协会成员国匈牙利盆景协会副会长
Attila Baumann Vice President of Hungarian Bonsai Association

首先，我很高兴参加2013"唐苑的世界盆景对话"国际年度论坛，虽然大家说着不同的语言，但通过盆景，我们聚在一起共同交流享受这美好的时刻。感谢苏放会长和中国唐苑苑主张小斌先生邀请我参加此次国际论坛。请允许我对我的国家匈牙利做个简短的介绍。匈牙利首都布达佩斯，面积93000km²，人口1000万。

匈牙利盆景文化最初的形成始于20世纪80年代，最初是汤玛斯·比罗在日本大使馆里看到的一本关于日本盆栽的宣传手册，汤玛斯·比罗阅读了这本宣传手册之后，撰写了一本有关盆景的英语读物，向匈牙利人介绍了一个英语盆景俱乐部。汤玛斯·比罗在这个俱乐部里结识了盆景界的朋友约翰·纳卡和保罗·莱斯尼维茨，并与他开始交往。这个俱乐部给我们提供了一个好机会让我们了解到盆中的树木如何生长，怎样对他们进行养护，使匈牙利人了解到盆景造型的基本形式。在匈牙利，我们面临的最大挑战有：怎样得到盆景盆？哪里买到专用工具、养料和适合栽培盆景的土壤？

匈牙利第一个盆景俱乐部成立于1983年，大学盆栽俱乐部，今年迎来了它的30周年庆典。

直到1990年，匈牙利的盆景艺术有些起色，有更多的盆景杂志和盆景俱乐部兴起。我们开始从荷兰、中国和德国进口盆景，从欧洲其他国家和日本购买并翻译有关盆栽的书籍和杂志，试图模仿制作原始形态的盆景作品，从而匈牙利的盆景艺术得到了进一步的发展。但仍然没有精美的盆钵和制作盆景的工具，盆景盆通常由混凝土和塑料制成的盆钵所代替，工具都是经过改造的适合盆景制作的工具。

20世纪匈牙利盆景有了更快的发展。在互联网上，我们有多个与盆景相关的博客、俱乐部网站和线上广告，很容易把大家聚在一起在上面各抒己见、相互切磋盆景技艺。其次我们有6个盆景俱乐部，3个在布达佩斯，3个在较小的城市，拥有了本土生产的盆钵，有200个盆景爱好者。其中不乏盆景艺术大师，他们亲自为树木造型，创造盆景作品参与国际展会并上台表演。我们有匈牙利语自主出版的盆景书籍。感谢以下诸位匈牙利盆景艺术大师为匈牙利的盆景艺术发展做出卓越的贡献。他们是大学盆栽俱乐部大师巴克索·安德拉斯，迪克·彼得，卡托纳·欧文，纳吉·阿帕德，帕普·桑德尔，拉奇·乔治。独立个人：布奇·约瑟夫，库斯·安德拉斯，西蒙·席萨巴。

匈牙利植被覆盖率很高，当地树种丰富，我们有：榆树（榆属）、橡树（栎属）、山毛榉（水青冈属）、山楂树（山楂属）、山茱萸（山茱萸属）、角树（鹅耳枥属）、紫杉（红豆杉属）、梨树（梨属）、枫树（栓皮槭属）、桤木（赤杨属）、灰树（白蜡树属）、白杨树（杨属）、菩提树（椴树属）、山梨树（花楸属）、云杉树（挪威云杉属）、松树（欧洲黑松，小林乌口树属）。

作为加入到欧洲盆栽协会的第20个成员国，我们的使命任重而道远，我们要努力促进匈牙利盆景文化的发展，支持匈牙利各地的盆景俱乐部、供应商以及促进盆景爱好者的艺术交流，并参与欧洲盆栽协会举办的盆景现场表演等活动，我们的任务是组织盆景展会、新星才艺大赛、参与国内和国际各大盆景展览、倡导盆景文化、与国家级和国际协会、俱乐部合作、

The Annual Forum 国际年度论坛

匈牙利盆栽协会举办的 2012 新星才艺大赛
Hungarian Bonsai Association held New Talent Contest 2012

匈牙利盆栽协会成立于 2011 年 6 月 11 日
Hungarian Bonsai Association was found on the 11th of June 2011

在全国各地挖掘盆景大师并提名。

最后我向大家展示近几年匈牙利人参与的国际性盆景大会。2012 年在卢布尔雅那，我们选送的一盆五针松盆景作品荣获优秀奖，伊娃·莱柯丝和安德拉斯·库斯获得水石奖。我们的新星巴林特·特尔派克也参加了本次大赛。2013 年 1 月匈牙利盆景爱好者参加了诺朗德斯杯盆景大赛。同年 3 月份我们选送理查德·蔡盖尔的 3 盆盆景作品参加在匈牙利欧丹库尔召开的欧洲盆栽大会展。

2012 匈牙利盆栽协会首次举办了新星才艺大赛。接着，在 2013 新星才艺大赛上我们制定了新的比赛规则。

2013 年 9 月匈牙利盆栽协会举办了首届盆栽展。展品繁多，有五针松，刺柏，和其他匈牙利当地树种，如梨树和橡树。一些盆钵也是出自匈牙利制盆大师费尔·佐尔坦和巴林特·特尔派克之手。

西蒙·席萨巴藏品 马齿苋树属 配盆：匈牙利 费尔·佐尔坦
Collector: Szimon Csaba. *Portulacaria*. Pot: File Zoltán Hungary

库斯·安德拉斯藏品 刺柏 *Juniperus formosana* 配盆：匈牙利 费尔·佐尔坦
Collector: Koós András. Taiwan Juniper. Pot: File Zoltán Hungary

匈牙利 拉奇·乔治吉藏品
Collector: Rácz Gyorgy Hungary

Hungarian
National Penjing Overview

Speaker: [Hungary] Attila Baumann Reorganizer: CP

I'm very glad to attend the Annual Forum of "the 2013 Dialogue to the World Penjing". Although we speak different languages, through Penjing we get together to make cultural exchange and enjoy the nice time. I sincerely thank you Mr. Su Fang and Mr. Zhang Xiaobin, the owner of China Tang Yuan for inviting me to join the forum. Please allow me to do a brief introduction about my country, Hungary. The capital of Hungary is Budapest, the area is 93,000km^2 and population is 10 million.

The first formation of the Hungarian Penjing culture began in the nineteen eighties with a brochure from the Japanese embassy, which was read by Tamás Bíró, and afterwards he published an English brochure which contained some contacts to an English Penjing club. Tamas Biró got in contact with Penjing friends in the English Penjing club, John Naka and Paul Lesniewicz. This was a good opportunity to let's understand how trees grow and how to maintain them in a small pot and the Hungarian people learnt the basic forms of Penjing creation. In Hungary we met big challenges, for instance, how to get a pot? Where to buy special tool, fertilizer and special soil?

The first club was established in 1983, University Bonsai Club, this club celebrated now the 30th anniversary.

Hungarian Penjing art hadn't some improvement until 1990. More Penjing magazines, publications and Penjing clubs appeared at that time. We began to import Penjing from Holland, China and Germany as well bought and translated some Penjing related books and magazines. We tried to imitate the original shape of Penjing, thereby Hungarian Penjing art obtain the further development, but still didn't have pots fitted for Penjing and tools, pots were mainly made from concrete, plastic, tools were created from reconstructed normal tools.

Let's talk about the current situation of Penjing in twentieth Century. We have many Penjing related blogs, club websites and internet presence, it's easy to get people together to let each one express his own view and learn from each other in term of Penjing techniques. Then we have 6 clubs in Hungary, 3 in Budapest, 3 in smaller cities, also we have own potteries, 200 Penjing enthusiasts in Hungary, there is no lack of masters among them. They personally make the modeling of trees, creat beautiful Penjing works, participate in international exhibition and demonstrate on the stage. We have own publications in Hungarian. I thank you all of Hungarian Penjing masters below for making outstanding contributions for the development of Hungarian Penjing art. They are Bacsó András, Deák Péter, Katona Ervin, Nagy Árpád, Papp Sándor, Rácz György from University Bonsai Club and Burschl József, Koós András, Szimon Csaba as independent.

The vegetation coverage is very high, National Species are rich. We have elm(*Ulmus pumila*), oak(*Quercus*), beech(*Fagus*), hawthorn(*Crataegus*), cornel(*Cornus*), hornbeam(*Carpinus*), yew(*Taxus*), pear(*Pyrus*), maple(*Acer campestre*), alder(*Alnus glutinosa*), ash(*Fraxinus*), poplar(*Populus*), linden(*Tilia*), sorb(*Sorbus*), spruce(*Picea abies*), pines(*Pinus nigra*, *Sylvestris*).

Hungary joined to the European Bonsai Association as the 20th member country. Our mission is a long way to go, we should try hard to promote the development of Hungarian Penjing culture, support the Penjing clubs, suppliers and the art exchange of Penjing lovers' around in Hungary, as well participate in European activities in Penjing live-demonstration. Our tasks are to organize exhibitions and New Talent Contest, to take part in national and international exhibitions, to advocate, to cooperate with national and international associations and clubs, to make the nomination of Penjing masters.

Finally, I want to show the recent international Penjing conventions Hungarian people participated in. In 2012 we participated in Ljubjana, where we won the prize "Award of Merit" with a *Pinus parviflora*, and the prize with a suiseki from Éva Lelkes and András Koós. Our new talent Bálint Tirpák participated in the Contest. In January 2013 Hungarian enthusiasts participated in *Noelanders Trophyand* in March we participatedon the EBA congress with 3 trees of Richárd Czégel from Hungary in Audincourt.

Hungarian Bonsai Association held its own first New Talent Contest in 2012. We have newly defined and written rules in New Talent Contest 2013.

The Hungarian Bonsai Association held its own first Bonsai Exhibition in September 2013. There were wide variety of exhibits, such as Japanese White Pine (*Pinus parviflora*), Chinese Juniper (*Juniperuschinensis* var. *sargentii*) and other Hungarian species like pear and oak. Some pots were manufactured by Hungarian potters Zoltán File and Bálint Tirpák.

匈牙利 马克兹卡盆栽工作室的盆栽作品展示
Collector: Marczika Bonsai Studio Hungary

一直在进步
——越南盆景
Keep on Growing
—Vietnam Penjing

阮氏皇,越南盆景协会主席、亚太盆栽友谊联盟理事长
BNguyen Thi Hoang, the Chairwoman of Vietnam Bonsai Association; the President of Asia-Pacific Bonsai Friendship Federation

发言人:【越南】阮氏皇 图文整理:CP　Speaker:[Vietnam] Nguyen Thi Hoang　Reorganizer: CP

水梅 *Wrightia religiosa* 高165cm 阮氏心藏品
Water Jasmine. Height: 165cm. Collector: Nguyen Thanh Tam

金莲木 *Ochna integerrima* 越南平定省盆景协会藏品
Collector: Binh Dinh Province—Bonsai Association

我谨代表身在越南的爱好自然和盆栽的社团成员发表一些见解:

越南拥有着悠久的传统文化和历史。越南人热爱自然。很久以前我们便开始种植植物和花卉。几乎在每个人的家里,你都能看见植物、花卉、鸟和鱼……然而,我们在最近才注意到了盆栽,而盆栽的种植和欣赏是在近20~30年间才发展起来的。

在学习盆栽的最初阶段,由于盆栽艺术家受传统方式影响将盆栽制作为装饰品,而使越南式盆栽看起来仍是小型的传统装饰性植物。我们将我们的世界观和对人类生命的看法投入于传统装饰性植物的制作中……这类植物体现了古代人的哲学并且给我们盆栽艺术家带来了巨大的影响。

在经过与世界盆栽产生了相互影响的一段时期后,越南的盆栽团体在盆栽理念和制作中发生了一些令人瞩目的改变。

他们开始创作出拥有更好的外观和象征意义的盆栽。但在那时,人们偏爱于尽量矮株的盆栽,或将盆栽的造型制作得尽可能奇怪。人们认为那是盆栽的审美标准。在近10~15年间,由于求知的渴望,越南的盆栽制作风格出现了许多显著的变化。盆栽越来越像自然中的树木,因为自然才是所有美丽事物的造物之主。自然是极其变化多端的,自然中没有条条框框,没有规定好的轮廓或标准。越南式的盆栽有着一些令人印象深刻的进步,这些进步缩短了其与世界盆栽之间的距离。

越南的盆栽艺术有着很高的技术和艺术水平,因为越南的人们非常努力而富有耐心。为了创作出精致的样式,越南的盆栽艺术家对他们的作品投入了很多心力。然而我们暂时还没有很多杰出的作品。我们投入的时间还不足以创作出一个出类拔萃的盆栽。但是您可以相信在不远的将来,越南盆栽将会拥有伟大的作品。我们有着年轻艺术家的坚定力量,包括那些富有创造性的人们,他们不停的学习新事物并全心投入到每一件盆栽作品中去。这里有着丰富的、各种各样的树木资源。一些树种完全可以用来制作优秀的盆栽样式,例如:茉莉、五月茶、水梅和象橘等。我希望这些树种能满足全世界盆栽爱好者们富有创造性的盆栽创作。

在我的国家,每个城市或省份在农历新年假日里都会举办展会。如果您想要对越南盆栽有更多的了解,这都是极好的参访机会。我们一直都欢迎你的到来。在2015年,ABFF(亚太盆栽友谊联盟)将在越南举行。这将是达到此目的的最佳机会。

在我们国家并没有专门针对盆栽的杂志,因为通常盆栽都被当作是园艺的一部分。在一些当地俱乐部中,学习和经验交流是十分受欢迎的,当然也通过展会和活动的形式进行。

在对中国的盆景进行观察后,我们得出以下看法:

2013 唐苑的世界盆景对话
DIALOGUE TO THE WORLD PENJING

茉莉 *Desmodium unifoliatum*
高 65cm 陈胜藏品
Height: 65cm. Collector: Tran Thang

茉莉 *Desmodium unifoliatum*
长 80cm 陈胜藏品
Length: 80cm. Collector: Tran Thang

五月茶 *Antidesma acidum* 高 160cm 阮氏皇藏品
Height: 160cm. Collector: Nguyen Thi Hoang

中国是盆景的发源地，因此你们拥有着悠久的盆景历史。贵国的盆景通过神奇的场景、风格和独树一帜的文化特征给人们留下了深刻的印象。

你们的盆栽技巧非常之高，我们十分崇拜并希望能向你们学习。

我们参加这次会议，代表着那些渴望通过参访和学习更多经验来发展越南盆栽的人们。我谨代表越南的盆栽爱好者们祝愿本次大会顺利举行，并向各位传达盆栽精神。

水梅 *Wrightia religiosa* 长 70cm 胡志明藏品
Water Jasmine. Length: 70cm. Collector: Ho Chi Minh

象橘 *Limonia acidissima* 高 80cm
黎光永藏品
Height: 80cm. Collector: Lê Quang Vinh

茉莉 *Desmodium unifoliatum*
越南龙川市盆景协会藏品
Collector: Long Xuyen province - Bonsai Association

水梅 *Wrightia reilgiosa* 长 147cm 宽 80cm
阮海鹏藏品
Length: 147cm, Width: 80cm. Collector: Nguyen Hai Phong

The Annual Forum 国际年度论坛

Keep on Growing
-Vietnam Penjing

Speaker: [Vietnam] Nguyen Thi Hoang Reorganizer: CP

I represent the community of people who are interested in the nature and bonsai in Vietnam to present some:

Vietnam has a long traditional culture and history. Vietnamese love the nature. We have begun growing plants, flowers for a long time. Almost in every house, there are the present of plants, flowers, birds, fish…. However, Bonsai has just has our attention recently, and growing and enjoying Bonsai has just developed for the last recent 20 to 30 years.

At the first stage of learning Bonsai, the Vietnamese Bonsai had the appearance of the traditional decoration plants but in the small size because bonsai artists are affected by the traditional way to make the plants for decoration. In the traditional decoration plants, we apply our worldview and the thought of human lives into making the plants…. This kind of plants is a philosophy of ancient people, and it places a huge impact on the thought of our bonsai artists.

After a period of time to interact with Bonsai in the world, Bonsai community in Vietnam has had some noticeable changes in the way of thinking and making Bonsai.

They began creating much better bonsais in the appearance and image. But at that moment, there is a preference on making the bonsai as short as possible or producing the bonsai which is as strange as possible. They considered those as a beauty standard of bonsai. In the last recent 10~15 years, because of the desire to learn more, there are many remarkable changes in the bonsai producing style in Vietnam. Bonsais have looked more similar to trees in the nature as nature is the master of every beauty. Nature is extremely diverse, and there is no frame, outline or standard in the nature. Vietnamese bonsai has some impressive progresses which make it go nearer to bonsai in the world.

The bonsai art in Vietnam has the high level of technique and art because Vietnamese are hard-working and patient. Vietnamese bonsai artists put much effort in their work to create sophisticated specimens. However, currently, we do not have many outstanding products. There is not enough time to make a bonsai product shine. But you can believe that in a near future, Vietnam Bonsai will have great specimens. We have the strong young artist force including people who are creative, non-stop study new things, and put their heart in each bonsai product. There is an abundant tree resource with diverse species. There are some kinds of them which are qualified to make good bonsai specimens, for instance: Desmodium unifoliatum, Antidesma acidum, Wrightia religiosa, Limonia acidissima... I hope that these trees will satisfy the creative bonsai production of bonsai lovers all around the world.

In my country, there are exhibitions on Lunar New year holidays in every city or province. If you want to learn more about Vietnam Bonsai, those are great chances to visit us. We always welcome you. In 2015, the ABFF (Asia-Pacific Bonsai Friendship Federation) will be held in Vietnam. That will be a wonderful occasion for this purpose.

In our country, there has not been any magazine which specializes on Bonsai because bonsai is considered as a part of horticulture in general. Learning and experience exchanging are popular in some local clubs, and through the exhibition and conventions.

After observing Bonsai in China, we have some following opinions:

China is the origin of Bonsai, so you have a long bonsai history. Your bonsai impresses other by amazing image, style and distinguish culture features.

Your bonsai technique is so high that we admire it so much and desire to learn from you.

We attend the convention this time as people who are eager visit and learn more experience to develop the bonsai art in Vietnam. I represent the bonsai lovers in Vietnam to wish the convention succeed and spread the bonsai spirit to everyone.

越南盆景欣赏 胡志明盆景协会藏品
Vietnam Penjing appreciation. Collector: Ho Chi Minh Bonsai Association

2013 唐苑的世界盆景对话
DIALOGUE TO THE WORLD PENJING

马来西亚盆景的起源
The Origin of Malaysia Penjing

发言人：【马来西亚】蔡国华 图文整理：CP Speaker: [Malaysia] Chua Kok Hwa Reorganizer: CP

马来西亚盆栽的起源及原始特征：简洁的形态、强大的根系、苍老的姿态
Origin of Penjing in Malaysia and original features of Penjing: concise form, strong root system, vigorous posture

莫泽熙先生（中左）和陈隆先生（中右）对推动马来西亚盆景事业发展做出了重要的成绩
Center left: Late Mo Ze Xi Center right: Mr. Tan Leong Cho. The both made a great contribution to the development of Malaysian Penjing

　　马来西亚盆景的起源可以追溯到19世纪初。在那个时候，从中国到东南亚的中国移民被称为南洋或南海。他们带着他们的语言、文化和艺术来到马来西亚、新加坡、印度尼西亚、泰国、文莱、菲律宾等地方。我是第三代炎黄子孙，我的爷爷来自莆田（位于中国福建省东部），我的先父出生于马来西亚。

　　在马来西亚的盆景协会有助于通过组织盆景展会和比赛来促进当地的盆景艺术。盆景协会包括：马来西亚盆景和雅石协会（Malaysia Penjing & Suiseki Society）、马六甲盆栽协会（Melaka Bonsai Association）、关丹县盆栽协会（Kuantan Bonsai Society）、吉兰丹盆栽协会（Kelantan Bonsai Society）、霹雳州盆栽和雅石协会（Perak Bonsai & Suiseki Society）、沙捞越盆栽协会（Sarawak Bonsai Association）。

　　在第五届亚太盆栽董事会会议以及1999年在台北举行的雅石和中国旧陶器大会期间，马来西亚盆景和雅石协会在廖先生的领导下赢得了在2001年在马来西亚吉隆坡举办第六届亚太会议的权利。本次大会帮助了马来西亚进一步提高他们的盆栽标准以与其他国家相媲美。在2012年中旬，黄荣钻先生（盆栽俱乐部董事兼盆栽和赏石学院院长）组织了马来西亚国际盆栽和赏石展会和大赛并在马来西亚开启了国际盆栽俱乐部的新篇章。国际裁判组来自美国、中国台湾、印度尼西亚、委内瑞拉、菲律宾。前国际盆栽俱乐部主席罗伯特·肯平先生也是国际裁判组裁判之一。

　　马来西亚盆景的特性与其他国家类似，尤其是台湾，马来西亚盆景与之有着密切的联系。马来西亚盆景强调根和分枝的开发以及精细和整洁的风格，并且关注盆景的树龄。

　　我们会邀请国际裁判来参加各种盆景活动，裁判们将分享他们对当地盆景的经验和想法，对盆景进行演示并召开问答会议。这些工作对于改善马来西亚的盆景标准有很大帮助。参加国际盆景展会和大会，例如：日本的国风展和大观展，中国的中国鼎国家大展、台湾的华风展、亚太盆栽大会、国际盆栽俱乐部大会也有助于盆栽发展、友谊和网络。

　　马来西亚没有举行过全国的盆景展会，所以要想了解更多关于马来西亚盆景的信息，我建议去参观由马来西亚的不同盆景协会组织的盆景展会和比赛。盆景展会每年将在吉隆坡、怡保、槟城、关丹县和沙捞越举行。

　　网络形式的盆景媒体在马来西亚比较普遍，脸谱（facebook）和网站在马来西亚人中非常受欢迎，例如：盆栽苗圃和三丰盆景园拥有5000多名粉丝。

　　我认为中国盆景有两个独特的特性：首先，独特的岭南盆景技术强调创造自然盆景，并突出分枝技术和天然成分。盆栽的布局和设计非常自然、美丽，且包含了中国艺术文化元素。第二，山水盆景需要采用中国的自然景观。它需要各种各样的岩石做景观以体现自然景观，例如：黄山。需要认真地对岩石进行熟练的采集、修饰、组装，并对浅盆景盆进行艺术方面的安排。山水盆景展示悬崖、陡峰、山谷、小岛和其他景观。这些都是我的最爱。

　　在不久的将来，几乎每个国家的盆景都会达到类似高水平的盆景技术和风格。如果我们只是寻求盆景技术和风格，我

The Annual Forum 国际年度论坛

蔡国华 来西亚盆景雅石协会、BCI 国际盆栽俱乐部会员
Chua Kok Hwa, member of Malaysia Penjing & Suiseki Society and BCI (Bonsai Clubs International)

们只会模仿其他盆景作品。在这种情况下，大多数盆景看上去都会差不多，从而丧失盆景的独特性和创造性。

盆景是一件艺术品，必须具备诸多艺术元素，例如：绘画、陶瓷、雕塑等。我列举一些伟大的绘画艺术家，如：齐白石、张大千、毕加索、草间弥生、李曼峰、阿凡迪、勒迈耶、卡里尔·易卜拉欣等来展示卓越的艺术。他们的作品都有自己独特的个人风格和性格、创造力、感情、内涵、特殊技术，还有效果、色彩、活力、观点、比例。我希望某天我能在我们的树木盆景或盆栽中看到这些优点和特质。

1980 年，陈隆先生在马六甲组织了首届盆景展会。在展会中展出的大多数盆栽都是他亲手制作的作品，并吸引了许多马来西亚人的关注
In 1980, Mr. Tan organized the very first Penjing exhibition in Malacca. Most of the bonsai exhibited were his creation and it attracted a lot of interest in Malaysia

在成功举办第一届盆景展会后，陈隆先生又在 1982 年和 1984 年在吉隆坡（马来西亚首都）举办了他的盆栽展会，图为 1984 年的展会
With much success of the first penjing exhibition, he later exhibited his bonsai in Kuala Lumpur, capital city of Malaysia in 1982 and 1984

1977 年，莫泽熙的水梅盆景作品
Master Mo Ze Xi's Penjing Species: *Wrightia Religiosa* Year: 1977

陈隆先生，一位来自马来西亚马六甲的咖啡商人，对莫泽熙成功的盆景展会留下了非常深刻的印象。他和另外两个朋友黄亚兴先生和胡林明先生多次到新加坡拜访莫先生并交换盆景经验和技术技能。陈隆先生后来建立了一个盆栽苗圃并命名为水梅园，并成为促进马来西亚盆栽艺术和贸易的先锋者
Mr. Tan Leong Cho a coffee merchant in Malacca, Malaysia, was very impressed with Mr. Mo's bonsai exhibition success. He and two other friends, Mr. Wong Ah Heng and Mr. Wu Lim Ming made frequent trips to Singapore to visit Mr. Mo to exchange bonsai experiences and technical skills. Mr. Tan later setup a bonsai nursery under the name of Sui Mui Yuan and became one of the pioneers in promoting bonsai art and trade in Malaysia

1986 年，陈隆先生在吉隆坡耀全购物中心组织了第一届全国盆景大赛。他邀请了威廉姆先生和另外两个新加坡人作为本次大赛的裁判。莫泽熙先生出席了这个比赛，共有 100 多盆盆栽参与了该比赛。图为大赛第一名作品
In 1986, Mr. Tan organized the first national Penjing competition at Yow Chuan Plaza in Kuala Lumpur. He invited Mr. William Goh and two others Singaporeans to be the judges for this show. Mr. Mo Ze Xi attended this show. Over 100 bonsai were exhibited

20 世纪 80 年代末至 90 年代是马来西亚盆景发展的重要时期，在这段时间里，又建立了许多盆栽协会和许多盆景苗圃，除了本土物种（例如：水梅、寿娘子、鹅掌枫和香松）外，某些苗圃甚至从日本、中国台湾、印度尼西亚和菲律宾进口盆栽。
1980s~1990s is the important stages of Penjing adevelopment in Malaysia. During this time, many bonsai societies were established and many bonsai nurseries were setup and some nurseries even imported bonsai from Japan, China Taiwan, Indonesia, Philippines, beside the home grown species like *Wrightia Religiosa, Premna Obtusifolia, Acer Mono, Baekea frutescens*

The Origin of Malaysia Penjing

2013 唐苑的世界盆景对话
DIALOGUE TO THE WORLD PENJING

Speaker: [Malaysia] Chua Kok Hwa Reorganizer: CP

20世纪80年代的马来西亚盆景作品（水梅）
The Penjing of Malaysia during 1980s (*Wrightia religiosa*)

The origin of Malaysia Penjing dated back to the early 19th Century. The Chinese immigrated from China to South East Asia, known as Nanyang or South Sea at that time. They brought with them their language, culture, and arts to Malaysia, Singapore, Indonesia, Thailand, Brunei, Philippines, etc. I am a third generation of Chinese descendant and my great grandpa came from Pu Tian, Eastern Fujian Province, China and my late father was born in Malaysia.

The bonsai societies in Malaysia help to promote the bonsai arts in their own state by organizing bonsai exhibitions and competitions. The bonsai societies are:

Malaysia Penjing & Suiseki Society
Melaka Bonsai Association
Kuantan Bonsai Society
Kelantan Bonsai Society
Perak Bonsai & Suiseki Society
Sarawak Bonsai Association

During the Board Meeting of the 5th Asia-Pacific Bonsai, Suiseki, and the Chinese Old Pottery Convention in Taipei in 1999, Malaysia Penjing & Suiseki Society under the leadership of Mr. Liaw and his team won the bid to host the 6th Asia-Pacific Convention in Kuala Lumpur, Malaysia in 2001. This convention had helped Malaysia to further improve their bonsai standard comparable with other countries. In mid 2012, Mr. I S Ng, a Director of Bonsai Club International and President of Bonsai and Stone Academy organized a Malaysia International Bonsai & Stone Exhibition & Competition in conjunction with the launching of the Bonsai Club International Chapter in Malaysia. The panel of international judges came from America, Taiwan, Indonesia, Venezuela, and Philippines. The former Bonsai Club International President, Mr. Robert Kempinski was also among the panel of judges

Malaysia Penjing features are similar to other countries especially Taiwan in which Malaysian Penjing have close contact with. Malaysia Penjing emphasizes on root and branch development, fine and neat styles, and focus on the age of bonsai.

International judges are invited to various Penjing events in Malaysia. Judges will shares their experiences and thoughts on local Penjing, they give demonstrations on bonsai, and Q & A sessions. These have helped improved the bonsai standard in Malaysia. Attending international bonsai exhibitions and conventions such as Kokufu, Taikan Ten In Japan, China Ding in China, Hua Feng Zhan in Taiwan, Asia-Pacific Bonsai Conventions, Bonsai Club International convention, etc. have also helped in bonsai development, friendship and network.

Malaysia does not hold a national bonsai exhibition. Therefore, there is no one particular bonsai exhibition that will give a full picture of Penjing in Malaysia. To know more about Penjing in Malaysia, I

The Annual Forum 国际年度论坛

20 世纪 90 年代的马来西亚盆景作品（海芙蓉、水梅、香松）
The Penjing of Malaysia during 1990s (*Atriplex maximowicziana* Makino, *Wrightia religiosa*, *Baecakea frustesens*)

2005 年的盆景作品（海芙蓉、寿娘子、水梅、真柏）
The Penjing work in 2005 (*Atriplex maximowicziana* Makino, *Premna obtusifolia*, *Wrightia religiosa*, *Juniperus chinesesis*)

2007 年马来西亚的盆景
The Penjing of Malaysia in 2007

Entering Bonsai Art
in Treasure Island, Taiwan

2007年马来西亚的盆景
The Penjing of Malaysia

suggest to visit Penjing exhibitions and competitions organize by the bonsai societies in various states of Malaysia. Bonsai exhibitions are held on yearly basis in Kuala Lumpur, Ipoh, Penang, Kuantan, and Sarawak.

Facebook and websites are very popular amongst Malaysians. For example, a bonsai nursery, San Fong（三丰盆景园）(https://www.facebook.com/sanfongbonsai) has more than 5,000 followers.

There are two unique features of Chinese Penjing :

First, the unique Lingnan Penjing technique which emphasizes on creating bonsai naturally, highlighting branches techniques, and natural composition. Bonsai layout and design is very natural, beautiful, and contains elements of Chinese art culture.

Second, the landscape bonsai（山水盆景）takes scenes from natural landscape in China. It requires a wide variety of rocks to do a landscape in order to reflect the natural landscape scenes, for example, Huang Shan（黄山）.Rocks need to be carefully and skillfully cut out and carve out, put together and arrange artistically in the shallow pot. Landscape bonsai show cliffs, steep peak, valley, island, and other landscape scenes. These are my favorites.

In the near future, almost every country's Penjing would have reached similar high level of Penjing techniques and style. If we are only seeking bonsai techniques and style, we will only imitate other bonsai works. In this case, most bonsai work will then look more or less the same. Uniqueness and Creativity are lost.

Bonsai is a work of art. Bonsai must possess elements of fine arts like paintings, pottery, sculpture, etc. Just to name a few great paintings artists like Qi Baishi（齐白石）, Zhang Daqian（张大千）, Piccaso, Yayoi Kusama, Lee Man Fong（李曼峰）, Affandi, Le Mayeur, Khalil Ibrahim, etc. demonstrate artistic excellence. Each of their paintings has their own unique personal style and character, creativity, lyrical, meaning, special technique and effects, use of colors, dynamism, perspective, and proportion. These qualities are ones in which I hope to see in our shu mu（树木）penjing or bonsai one day.

Thank you, and have a pleasant day!

2009年马来西亚的盆景
The Penjing of Malaysia in 2009

The Annual Forum 国际年度论坛

2011 年马来西亚的盆景
The Penjing of Malaysia in 2011

2012 年马来西亚的盆景
The Penjing of Malaysia in 2012

马来西亚的高水平盆景作品
High standard Penjing of Malaysia

一些水平比较好的盆景
Some good Penjing

捷克共和国的盆景进化过程
Process of Penjing Evolution in the Czech Republic

发言人：【捷克】斯瓦托普卢克·马特杰卡　图文整理：CP
Speaker: [the Czech Republic] Svatopluk Matejka　Reorganizer: CP

斯瓦托普卢克·马特杰卡《盆栽》杂志主编
Svatopluk Matejka Redactor of *Bonsai* Magazine

首先，我向大家介绍捷克共和国盆景发展的初始阶段。

20世纪80年代开始大规模引进盆景。当时，布拉格盆景俱乐部很有影响力，它是由我国的盆景界先驱兹德内克·赫尔德利奇卡及妻子维拉·佩特·赫瑞耐克、卡雷尔·斯托尔克和其他人一起创办的，现有会员超过1200名。它发行杂志，出版了第一本盆景方面的书籍，组织了非常热门的展会，并首次引进了中国的盆景盆。针对盆景的栽培，他们还努力激发人们对不为人知的植物产生兴趣，几乎每个月他们都会为来自全国各地的成员组织讲座。此外，那些在盆景方面有经验的人也很乐意与他人分享。此后，人们在不同的城镇成立了其他盆景俱乐部，这些俱乐部里大师云集，有经验丰富的树木学家、园艺大师和高山植物专家。但是，主要问题仍然是树木的造型问题。捷克国内没有任何现成的关于盆景制作步骤、造型技法等方面的知识，也没有盆景美学的理论阐述，同时还没有可用的盆景盆、盆景土壤、工具甚至铝线。一般都是通过简单的修剪来处理盆景。

1989年，也就是发生在天鹅绒革命（西欧边界向捷克人开放）之后，我们的盆景取得了突破性进展。我国制作盆景的技术和知识储备已经迅速赶上了西欧盆景栽培者当时的水平。在那个时刻，敏锐的利博霍维采盆景俱乐部组织了短途旅行去拜访主要的西欧盆景栽培大师和著名的盆景中心：海德堡（德国）、洛德（荷兰）以及辛兹纳赫（瑞士）。他们了解到了新的造型方法，并且也采购到所需的材料。1997年，瓦茨拉夫·诺瓦克和斯瓦托普卢克·马特杰卡借着举行利博霍维采盆景俱乐部的年度会议之际成立了捷克盆景协会，并邀请了所有著名的捷克俱乐部的代表。新协会为各俱乐部与国际盆景的合作架起了一座桥梁，代表捷克与外国联系并作为我国盆景主流的倡导者出席世界盆景活动。自1998年起，捷克盆景协会发行了自己的杂志，并成为欧洲盆景协会成员中第一个东欧国家。此外，捷克共和国还拿出了10个展品参加了2001年在慕尼黑召开的全球性的第六届新星才艺大赛。历经15年的探索，我国盆景的发展主要体现在：与世界上其他盆景事业较成熟、发达的国家积极地互动，定期参与由欧洲盆栽协会举办的欧洲盆栽大会和组织本国每年一届的国家展会。由于一些捷克盆景制作者跟国外的盆景大师实习，捷克已经融入整个欧洲的盆景大环境，同时水石艺术的发展与之也是齐头并进的。

在我国盆景界，我非常尊重以下各位盆景大师：瓦茨拉夫·诺瓦克（捷克盆景协会会长），杰纳·卜辞，帕维尔·斯洛瓦克，卢卡斯·思若特尼，简·库莱皮，米兰·卡皮塞克，奥勒达·卡斯帕，迈克尔·布迪克，马丁·阿森布伦纳，黎博·卡杰斯还有我本人，斯瓦托普卢克·马特杰卡。这些人都曾在重大赛事上数次获奖。除此之外，他们还在新星才艺大赛中取得过名次，作品也在享有声望的国际大展中展出过。

我认为我们国家用来做盆景的桩材主要来自欧洲的树种，这是捷克盆景作品中显著的特点，在大自然中寻找创作灵感。

斯瓦托普卢克·马特杰卡先生在自家盆景园中劳作
Svatopluk Matejka is working at his own Penjing garden

斯瓦托普卢克·马特杰卡先生正在对盆景进行日常的养护
Svatopluk Matejka is doing daily maintenance for Penjing

The Annual Forum 国际年度论坛

左图: 斯瓦托普卢克·马特杰卡先生（左）和瓦茨拉夫·诺瓦克（右）正在制作盆景
Left picture: Svatopluk Matejka (left) and Václav Novák (right) are making Penjing
右图: 捷克盆景展示
Right piocture: Penjing exhibits in the Czech Republic

如果有人有兴趣了解捷克盆景美学，我建议去参观每年在不同乡镇举行的国家盆景展会和水石展。我们也开展一些关于盆景的研讨会和讲座，例如：有关东方花园的讲座。我们的这些活动通过两本捷克杂志进行推广：第一本名为《盆栽》杂志，由捷克盆景协会出版；第二本名为《盆景和日本花园》杂志。

我们将中国视为盆景的摇篮。就我们所了解的中国盆景而言，我们钦佩中国盆景呈现出的自然形态美以及意境的表现手法高于形式美。在我看来，他们的自然主义是最主要的一点。我们可以看到中国盆景不像一棵单独的树木，而更像自然的一部分。我喜欢盆景通过各式各样的形态突显自然气息。

对于盆景的发展前景我很有信心。我在欧洲展会上有时会看到不寻常的、创新的技法，奇特形状的盆景盆，意想不到的外国树种制作的盆景作品。我真的认为盆景发展趋势将是回归自然，所以大多数盆景都展示了它们的本质特征。因此，盆景的表达将会是最真实和最自然的。

First I'd like to talk about the initial stage of Penjing evolution in the Czech Republic.

The beginning of mass introduction to Penjing happened in the 80's of the 20th century. At that time the influence of Prague Penjing club is noticeable, this club was founded by the early pioneers of this art in our country: Zdeněk Hrdlička, his wife Věra, Petr Herynek, Karel Štolc and others, this club had over 1200 members. It issued its magazine, published first books about Penjing, organized very people to visit exhibitions, arranged the first import of Chinese pots, for Penjing cultivation, they stimulated people's interests for public unknown plants and also organized lectures for members from whole country almost each month. Those who had some experiences with Penjing shared them with others. Thereafter other clubs were founded in different towns. They were associated around experienced arborists, gardeners and experts on alpine plants. The main problem remained the question of shaping trees. There weren't available any knowledge of procedures and forming methods, aesthetics of Penjing wasn't elaborated, pots, potting soils, tools even wires weren't available. Penjing were treated especially by cutting.

Breakthrough occurred in 1989, after the velvet revolution when the borders of Western Europe were opened to Czech people. The Czech Republic quickly caught up with what Western European Penjing culturists already knew. In that time agile Libochovice Penjing Club organized excursions to visit major Western European Penjing culturists and to significant Penjing centers: Heidelberg (Germany), Loder (Netherlands), Schinznach (Switzerland). Participant of these meetings met the methods of shaping and also could buy the needed material. In 1997 Václav Novák and Svatopluk Matějka founded the Czech Penjing Association on the occasion of annual meeting of Libochovice Penjing Club, where the representatives from all known Czech clubs were invited. The new association is arching over the work of clubs, represents Czech scene in contact with foreign countries and is an organizer of Penjing stream in our country. Since 1998 the Czech Penjing Association has issued its own magazine and became a member of European Penjing Association as the first country from Eastern Europe. The Czech Republic sent ten exhibits and take part in competition of new talents in the VI. Worldwide Exhibition in Munchen in 2001, in this new 15 years long era —— characterised by active contact with grown up Penjing world, with European and worldwide respected Penjing masters, by regular participation in EBA congresses held by EBA congress organization, by organization of own annual national exhibition. Thanks to participation of some Czech Penjing culturists in internship abroad with Penjing masters–Czech gets into context of European Penjing community, it's the same in suiseki art.

At our Penjing circle I fully respect these Penjing culturists: Václav Novák– the president of Czech Penjing Association, Jana Bučil, Pavel Slovák, Lukáš Sirotný, Jan Culek, Milan Karpíšek, Oldřich Kašpar, Michal Budík, Martin Ašenbrener, Libor Kajš and me –Svatopluk Matějka. These people were awarded several times, they won in the New Talent competition, and their works were exhibited at prestigious abroad exhibitions.

Subsequently, I introduce the characteristic features of Czech Penjing activities. I consider what species form our timber species –European species, as the peculiar features of our Penjing activities. We are looking for inspiration in our nature. The originality is that we often use pots made in our Czech workshops, for example Studio Klika & Kuřátková, Isabelia, Dan Pánek etc.

If someone is interested in getting to know Czech Penjing event, I would recommend to visit national Penjing exhibition and suiseki CBA, that takes place once a year in different towns. Many workshops and lectures also take place in, for example about oriental gardens. These activities are promoted especially by these two Czech magazines: *Bonsai* (magazine of Czech Penjing Association) and *Penjing and Japanese gardens*.

We respect China as the cradle of Penjing. China Penjing as far as we know, we admire their natural free look, where expression prevails over form. Personally, their naturalism is the most valuable point. And moreover we can see China Penjing not like a single tree but like a part of nature. I like emphasizing of natural atmosphere by various figures.

It is too confident for me to guess the development of Penjing. I just noticed that on European exhibitions sometimes appear unusual avant-garde approaches, unusual shapes of pots, unexpected foreign materials built into Penjing composition. I really want to think that the trend in Penjing development will be their comeback to natural look, so that most resemble tree in nature. So the express of Penjing would be truest and most natural.

2013 唐苑的世界盆景对话
DIALOGUE TO THE WORLD PENJING

至爱盆景
Loved Penjing

发言人：【瑞典】玛利亚·阿尔博尔莉思·罗斯伯格 图文整理：CP
Speaker: [Sweden] Maria Arborelius-Rosberg Reorganizer: CP

瑞典地理位置
The location of Sweden

驯鹿和圣诞老人
Reindeer and Santa Claus

瑞典优美的自然环境
The beautiful scenery of Sweden

马尔默的美丽风光
The beautiful scenery of Malmo

尊敬的各位来宾，你们好。我的名字是玛丽亚·阿尔博尔莉思·罗斯伯格，我来自瑞典。瑞典位于欧洲北部，是斯堪的纳维亚半岛的组成部分，有950万居民。

瑞典位于北极圈北部以"午夜太阳"的奇景而闻名，夏季时，始终都是太阳普照，什么时候都是白天！当然，在冬季就完全相反了，白天只有几个小时，天气非常寒冷且覆盖着很多积雪。

瑞典的首都是斯德哥尔摩，位于国家的中部以东部分。它是一个位于群岛环绕的美丽地方。瑞典是君主立宪制国家，所以有住在城堡中的国王和王后。瑞典北部地区有许多驯鹿，我们相信圣诞老人的家就在那。

瑞典南部地区（也就是我的家乡）那里有一个更温和的气候和最适合发展农业的地区。我住在瑞典第三大城市——马尔默，这里居住着30万位居民。自2000年以来，这里建成了一座宏伟的桥梁，它连接着马尔默和哥本哈根（丹麦首都）。这两个国家有许多共同之处，很多人都从马尔默到大城市哥本哈根寻找工作，因为那里的机会更多。

马尔默冬季的温度很少低于零下10～15℃，夏季温度为20～25℃。利于植物生长的季节非常短，所以对盆景的生长也是一种阻碍，一年中并没有一个很长的增长期。树木从4～5月份开始长出新的叶子，但是到7～8月，它们就停止了生长。因此，我们必须非常有耐心！

我从大约30年前开始研究盆栽。我在杂志上读到有关盆栽的文章，出于好奇，我认为我可以从我父母的花园中选取材料来制作我自己的盆栽。虽然我对盆栽一无所知，但是我却准备从枸子属植物和栗子树种开始我的盆栽制作。我的许多实验性的盆栽作品都没有存活，但我却积累了一些经验，虽然这些经验并不充足。直到我发现瑞典盆栽协会并成为其中一员，关于盆栽方面的知识越来越多。我每周末都会去参加了一个适合初学者的课程，在那里我真正收获到了很多关于盆栽的知识。现在我的那些毫不起眼的小树木有了一个更大的存活几率，我的爱好给我带来了很多乐趣。

从那以后，我继续使用不很昂贵的材料来制作盆栽，并且收藏了许多瑞典本地树木制作的盆栽。拥有了更多关于盆景方面的知识，当然我也购买了一些进口的盆栽。位于欧洲的荷兰是一个从远东进口大量植物的国家，同时它也是一个著名的花园植物的栽培基地，尤其是鳞茎植物，例如：郁金香和水仙花。我也从荷兰购买了漂亮的皋月杜鹃花、日本槭树和五针松。

在瑞典，像许多其他国家一样，当人们谈到盆栽时，他们便会想到日本。他们试图将他们的盆景制作成像日本那样的古典形式，他们采用的物种也都是日本槭树和松树。但是在我们国家，由于我们的气候原因导致我们的盆景看起来完全不同。因此我们最好试图将旧传统转移到我们本地树种，并在将来发展这一传统。我们使我们国家的盆栽看起来有瑞典的风格，但是又不失传统。

来自北半球的树木包括：疣皮桦、挪威云杉、小叶椴、栓皮枫、英国栎、欧洲山毛榉、鹅耳枥、光榆、欧洲七叶树、落叶松、北欧红松等。

我真的非常荣幸来到中国欣赏之前我一无所知的优秀的中国盆景。我只在很多年前从一本名为《今日盆栽》的杂志上刊登的文章中听说了漂亮的中国盆景，但是现在这本杂志已经停刊了。现在在欧洲，我们主要是从一本名为《盆栽聚焦》的荷兰杂志上来获取盆栽信息。对所有欧洲的盆栽爱好者而言，它是一个非常好的信息来源。

The Annual Forum 国际年度论坛

玛利亚·阿尔博尔莉思·罗斯伯格，欧洲盆栽协会成员国瑞典盆景协会代表；瑞典盆景协会理事会成员。
Maria Arborelius-Rosberg, the delegate of Sweden in the European Bonsai Association; board member of the Swedish Bonsai Association

我非常高兴有这个荣幸去学习很多新的东西，并将这些盆景知识带回到我的国家。在我来这之前，我只做了一个非常小的尝试：挪威云杉、圆柏和假山毛榉。

下面我将着重讲解这次议会的主要议题：

瑞典的盆景历史很短。瑞典盆景协会作为一个非盈利团体成立于1986年。从20世纪70年代开始，有人开始使用水蜡树制作盆栽。

瑞典盆景协会的宗旨是提高人们对于盆景的兴趣，并深入地理解瑞典盆景，并提高其成员的盆栽栽培知识。遗憾的是，我们现在只有大约130位会员，但是10年前我们却有400名会员。

当地会议一般都在瑞典的四个主要地区进行，旨在让成员开心地交流经验。此外还有一个年度会议，通常邀请国外的盆栽大师进行示范、讲座及研讨会。与此同时，我们也有一个全国性的展会。今年的展会我安排在马尔默举行。

互联网是一个用于访问我们主页和论坛的最受欢迎的方式。这也可能是我们为什么失去那么多成员的原因，许多年轻人可以通过互联网得到很多的信息而无需支付任何费用。我们的预算非常有限，因此无法进行大型活动。我们能做的仅是发布杂志和租用一个地方来举行会议。

瑞典盆景协会的杂志是季度刊。从2001年开始，瑞典盆景协会也成为了欧洲盆栽协会的成员。

因为在瑞典我们没有任何人专门从事盆栽和盆景行业，所以没有什么代表性人物。我们是爱好盆景的业余爱好者！随着时间的推移，我们现在具备了更多盆景知识，一些盆景作品已经具备了成为杰作的潜力。在瑞典，我们有泥炭沼泽和矮化松树可以作为天然的盆栽作品，在某些情况下这些树的历史多达上百年！

唯一的大型展会就是之前提到的一年一度的会议。

因为我之前对中国盆景毫不了解，因此我必须提前掌握一些中国盆景的知识。我只知道它更多的是关于山水思想并且会有一些石头作为配饰。2007年，我和我的家人到中国游玩，度过了一个完美的假期。我们在上海、苏州和杭州看到一些非常美丽的花园。在桂林，我明白了为什么桂林美丽的风景能代表艺术。它是如此的奇妙！当我们欣赏盆景时我们便能看到中国的山水画。

我希望我们国家能发现盆栽或盆景的乐趣，并视其为一个有益的爱好。许多现代人都在强调，这个爱好能让人感到放松并平静下来。我们努力将自己的传统与当地树种结合起来一起发展，希望从中国盆景中找到灵感并将其用运用到制作我的瑞典盆景，使它变得更有"中国味"。

瑞典盆景协会主页
The webpage of Swedish Bonsai Association

瑞典盆景协会所出版的刊物
The publication of Swedish Bonsai Association

瑞典盆景协会所举办的展览活动
The exhibition held by Swedish Bonsai Association

瑞典盆景协会所举办的展览活动
The exhibition held by Swedish Bonsai Association

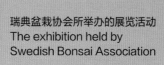

瑞典盆景协会所举办的展览活动
The exhibition held by Swedish Bonsai Association

Loved Penjing

Speaker: [Sweden] Maria Arborelius-Rosberg Reorganiz[er]

瑞典盆栽协会所举办的展览活动上的精美展品 杜鹃 高50cm 宽60cm
玛利亚·阿尔博尔莉思·罗斯伯格藏品
The fine exhibits in the exhibition held by Swedish Bonsai Association.
Azalea. Height:50cm, Width:60cm. Collector:Maria Arborelius-Rosberg

My name is Maria Arborelius Rosberg and I come from Sweden. It is a country in the north of Europe. Our country is a part of Scandinavia. We have 9.5 million inhabitants.

Sweden is a country well known for "the midnight sun" in the northern part at the Polar circle. In the summer the sun shines all night! Of course it is the opposite in the winter, only a few hours of daylight and very cold weather with lots of snow.

Our capital is Stockholm, in the middle-east part of the country. It is a beautiful city situated on many islands with a large archipelago. We have a king and queen that live in a castle. In the north we have reindeer and we believe that Santa Claus lives there.

In the southern part of the country, where I come from, we have a much milder climate and the best agricultural district of the land. I live in Malmo, Sweden's third largest city, with 30 thousand inhabitants. Since the year 2000 we have a beautiful bridge that unites Malmo and Copenhagen, the capital of Denmark. The two countries have a lot in common and a lot of people from Malmo work in Copenhagen, since there is more work in the bigger city.

In Malmo the winter it is seldom colder than minus 10 ~ 15 degrees Celsius and in summer the temperature is 20 ~ 25 degrees Celsius. The growing season is rather short and it is bad for growing Penjing since in one year we have not a big growth. The trees start to have new leaves in the month of April, May and already in July, August we have no more growth. We have to be very patient!

I started with Bonsai as a student about 30 years ago. I read about Bonsai in a magazine and by curiosity I thought that it could be possible to create my own bonsai from material in my parents' garden. I did not know anything, but started with small Cotoneasters and Chestnuts. A lot of my experiments did not survive, but I learned some by doing, but not a lot. Not until I found out about the Swedish Bonsai Association, and joined in as a member, things got better. I went on a week-end course for beginners and there I had a true revelation: now my humble little trees had a better chance to survive and all the fun with this hobby came to me.

Since then I have continued to make my own Bonsai from inexpensive material and I now have a collection of Bonsai from native trees from Sweden. With more knowledge I of course also have bought imported Bonsai. In Europe the Netherlands is a country which imports a lot of plants from the Far East and also is famous for cultivation of garden plants and especially bulbs, as tulips and narcissus. I have bought wonderful Satsuki Azaleas, Japanese Acers and Pinus parviflora in the Netherlands.

In Sweden, like in many other countries, when people think of Bonsai, they think of Japan. They try to form their trees like the classical forms in Japan, and the species are Japanese Acers and Pine trees, but in our country with our climate, we have trees that look totally different. Therefore it is better that we try to transfer the old tradition to our native trees and develop this tradition into the future. We make the bonsai look like the trees look in our country but with all the tradition as well.

Trees from the northern hemisphere are: Betula verrucosa, Picea abies, Tilia cordata, Acer campestre, Quercus robur, Fagus sylvatica, Carpinus betulus, Ulmus glabra, Aesculus hippocastanum, Larix,

演讲人玛利亚女士的盆景藏品
Penjing collection of Maria Arborelius-Rosberg

瑞典盆栽协会所举办的展览活动上的精美展品 长白松 *Pinus sylvestris*
高100cm 本特·斯托尔茨藏品
The fine exhibits in the exhibition held by Swedish Bonsai Association.
Mongolian Scots Pine. Height:100cm. Collector: Bengt Stoltz

The Annual Forum 国际年度论坛

栓皮槭 *Acer campestre*
高 20cm 宽 20cm 玛利亚·阿尔博尔莉思·罗斯伯格藏品
Field maple. Height:20cm, Width:20cm. Collector:Maria Arborelius-Rosberg

鸡爪枫 *Acer palmatum* 高 15cm 宽 20cm 玛利亚·阿尔博尔莉思·罗斯伯格藏品
Japanese Maple. Height:15cm, Width:20cm. Collector:Maria Arborelius-Rosberg

鸡爪枫"出猩猩" *Acer palmatum* 高 40cm 宽 45cm 玛利亚·阿尔博尔莉思·罗斯伯格藏品
Japanese Maple. Height:40 cm, Width:45cm. Collector: Maria Arborelius-Rosberg

欧洲七树 *Aesculus hippocastanum* 高 30cm 宽 20cm 玛利亚·阿尔博尔莉思·罗斯伯格藏品
Common Horse-Chestnut. Height: 30cm, Width:20cm. Collector:Maria Arborelius-Rosberg

平枝栒子 *Cotoneaster horizontalis* 高 50cm 宽 30cm 玛利亚·阿尔博尔莉思·罗斯伯格藏品
Rock Cotoneaster. Height:50cm, Width:30cm. Collector:Maria Arborelius-Rosberg

欧洲山毛榉 *Fagus sylvaticus* 玛利亚·阿尔博尔莉思·罗斯伯格藏品
European Beech.Collector: Maria Arborelius-Rosberg

倭海棠 *Chaenomeles japonica* 高 25cm 宽 30cm 玛利亚·阿尔博尔莉思·罗斯伯格藏品
Dwarf Flowering Quince. Hight:25cm, Width:30cm. Collector: Maria Arborelius-Rosberg

Pinus silvestris, etc.

For me it is a great privilege to come to China and to be able to see all the wonders of Chinese Penjing that I did not know anything of before. I have only learnt of the wonderful Penjing from China from an article written in the Bonsai Today magazine many years ago. This magazine does not exist anymore. In Europe now we have the Bonsai Focus magazine from the Netherlands as our main source of Bonsai information. It is an excellent source of information for all bonsai lovers in Europe.

I am very happy to have the great honor to learn a lot of new things about Penjing and to take this knowledge back home to my country. I have only made one humble attempt before I came here, with a Picea abies, Chinese juniper and a Nothophagus Antarctica.

Reference to your topics:

The history of Penjing in Sweden is very short. The Swedish Bonsai Association was founded in 1986 as a non-profit making association. There had been people working with bonsai probably in privet since the 1970s.

The aim of the association is to increase the interest and understanding of bonsai in Sweden and to improve the knowledge of bonsai cultivation amongst its members. Sadly we have only some 130 members at the moment, when we 10 years ago had 400 members.

Local meetings are held around Sweden in four main districts were members exchange experience and have fun together. There is also an annual meeting where usually a Bonsai master from abroad is invited to hold demonstrations, lectures and workshops. At this occasion we also have a national exhibition. This year I arranged this event in Malmo.

The internet is a popular place to visit our homepage and the discussion forum. This may also be the reason why we lose so many members, many young people can get a lot of information without having to pay the fee of membership. We have a very modest budget and cannot afford big events. Just to pay the cost to hire a place to hold meetings is what we can afford at the time and for publishing our magazine.

The society magazine comes out quarterly. We are also a member of the European Bonsai Association since 2001.

Since we in Sweden not have anyone working with bonsai/ Penjing as professionals we cannot point out anyone as special representative. We are glad amateurs! With time passing we are now starting to have greater knowledge and there are some trees that are starting to have potentials as masterpieces. Specially, trees of Yamadori origins. In Sweden we have peat bogs and there are wonderful natural dwarfed pines that are natural bonsai, in some cases up to a hundred years old!

The only big exhibition is the annual meeting as mentioned.

I have to learn about the China Penjing since I have no knowledge. I only know that it is more about landscapes and a lot more stones in Penjing have visited China in 2007 as a tourist with my family and we had a wonderful vacation all around China. We saw some very beautiful gardens in Shanghai, Suzhou and Hangzhou. In Guilin I understood why the beautiful scenery is represented in the arts. It is so fantastic! A representation of the Chinese landscape is what we see when we look at Penjing.

I hope that more people in my country will find the joy of having bonsai/Penjing as a rewarding hobby. Many modern people are stressed and this hobby is very relaxing and calming. We will try to develop our own tradition with our native trees and now from being in China I hope to take back with me inspiration to make my Swedish Penjing a little more "Chinese"

Thank you.

友谊之桥——盆栽
The Bridge of Friendship — Bonsai

2013 唐苑的世界盆景对话
DIALOGUE TO THE WORLD PENJING

发言人：【丹麦】汤米·尼尔森　图文整理：CP　Speaker: [Denmark] Tommy Niesen　Reorganizer: CP

丹麦盆景欣赏
Denmark Penjing appreciation

大家好！

有些人可能认识我，但大多数人并不认识我。请允许我先做一下自我介绍。

我的名字是汤米·尼尔森（Tommy Nielsen），我是丹麦盆栽协会的会长。三年前，当选会长，是协会有史以来最年轻的会长。今年31岁，已婚。我可爱的妻子名叫莫妮卡（Monica），我们住在丹麦一个叫做道格比耶格（Daugbjerg）的小镇。道格比耶格（Daugbjerg）非常小，只有600人。这儿没有大城市的喧嚣，只有鸟儿、狐狸和鹿的叫声。

丹麦盆栽理事会是一个只有32年历史的新生协会。每年我们都会在4月或5月任选一个周末举办活动，活动中我们会评选出最优秀的盆栽和小品组合盆栽。虽然我没有任何作品参展，但我每年都很期待这一天的到来。在活动中，你不仅能与老朋友见面，还能认识新的朋友，你在欣赏其他人的作品的同时，还能从中得到启发。

丹麦盆栽协会成立时拥有将近700个会员，人们自发带着他们的素材相聚在一起并互相探讨交流。协会经过多年的不断壮大，到20世纪90年代初，丹麦已有780人从事盆栽。此后，协会的成员越来越少，至今我们只是一个有着253个成员的小协会。

哪儿出错了吗？

会员大多是老年人，而90年代的老会员很多都已离世。盆栽变成了少数人的爱好。年轻人不明白盆栽的迷人之处，更不能理解这是一个多么完美的爱好。而且，在他们没有时间倾听的时候，你也很难向他们诉说。还有一个问题，要使那些曾经年轻的树株成为如今非常强而有力的盆栽需要长时间的工作。我想这或许也是一些人开始畏惧进行盆栽制作的原因。

当我被邀请进行这次关于丹麦盆栽文化的演讲时，我想到了一个问题：为什么你们要制作盆栽呢？

我9岁时，在丹麦的一个日式花园里初遇盆栽。我7岁开始练习空手道，从那时起我便迷恋于盆栽和亚洲文化。我不喜欢足球和其他团队运动，在学校也经常被人戏弄，于是我只能去道场练习。在我10岁的时候，我的父母得到了一棵盆栽。在我印象中，那并不是一棵生长旺盛的树，但我却深深地被它迷住了。因为我们不知道如何换盆和换土，那棵盆栽枯死了。不幸的是，从超市买回来这棵盆栽的时候没有说明书，所以它死了。当我18岁的时候，父母送给我一本名叫《盆栽入门》的书，我的盆栽生涯就此开始。那本书写的并不是很好，但它教会了我一些盆栽的简单知识。

我加入了当地的一个俱乐部，和一些年长的成员一起踏上了盆栽之路。最初，我看了许多电影。其中有一部是关于我的偶像凯文·威尔森（Kevin Wilson）使用重型工具创造杰作的。

The Annual Forum 国际年度论坛

汤米·尼尔森 欧洲盆景协会成员国丹麦盆景协会会长
Tommy Niesen
President of Danish Bonsai Association

我开始寻找大型的素材并开始进行创作，这需要大量的工作，这一过程让我感到轻松和享受。我将盆景创作作为减压和寻找内在平衡的方法，这是我想做能做的事情。我可以几个小时不间断地进行创作，而不用去配合其他人的计划。一段时间之后，我仍然可以在这一素材上进行新的创作。

七年的时光过去了，我开始尝试日本人的创作方法，试着去做同样的作品。此后，我的爱好产生了一个新的方向，莫滕·阿尔贝克（Morten Albaek）让我了解到小品盆栽，他告诉我小树中蕴藏着的秘密和其拥有的强大力量。莫滕给了我很多小品盆栽创作的启发，而托本·布兰德费尔德（Torben Brendfeldt）和伊冯·加贝克（Yvone Garbaek）则教会我如何养护这些盆栽。他们还教会了我盆栽展览中的知识，让我可以思考盆栽的真正含义。

2012年，我见到了瓦茨拉夫·诺瓦克（Vaclav Novak），他当时在我们的年度活动中进行现场创作表演和学术研讨。因为瓦茨拉夫也担任评委，因此我想知道他对不同奖项的看法和评选原因。他告诉我一种与日本盆栽不同的盆栽制作方式，盆栽创造需要自然，尽管这些素材不是山采的，但是也能给我们讲诉自然的故事。盆栽完全是关于创作和幻想，就像是描绘一幅图画或是书写一本故事书，它们应诉说着一个藏在树后的男孩或女孩的故事。这于我，是全新的，却让我从另外一个角度认识盆栽。

2013年，我见到了我的偶像之一凯文·威尔森（Kevin Willson）和他所携带的重型工具。那工具对我来说太大了，而凯文则可以随心所欲地使用它。凯文会说许多有趣的故事，我们成为了朋友。他还说他在活动中并没有给盆栽做评比，让我们自己来判断盆栽的好坏。凯文不想这样对盆栽进行评比，是因为他相信所有盆栽都是平等的，创作者的工作便是通过盆栽来诉说故事。凯文与瓦茨拉夫说了同样的话，但是他同时也认为人们在创作盆栽时应来源于自然而高于自然，应于创作者自己国家的树木形状来做为盆栽创作中的启发。

在过去的两年里，我一直在思考自己进行盆栽创作的原因。我并不是在考虑停止创作，而是在问自己为什么要创作盆栽。你可以尝试并向自己提问：为什么我要创作盆栽。对于我来说，找到这个答案并不容易，我必须一次又一次地重复问自己这个问题。最终，我找到了两个答案。

（1）通过创作盆栽，我可以表达自己的想法和展示自己的作品。在盆栽创作中，我可以修剪、换盆、蟠扎和再次造型，做我想做的任何事情。盆栽使我心情平和，放松自己并忘记那些带给我压力的事情。但是，当我们问自己为什么我们要创作盆栽时，最重要的是盆栽将我们聚集到了一起并跨越全世界各种界限建立起友谊。

（2）通过盆栽，年轻人和老人之间也能建立起友谊。这便是盆栽与众不同之处，您能想到有什么爱好能让一个15～20岁的年轻男孩或女孩与一个能当他们祖父或祖母的长辈保持着纯真的友谊呢？在我看来，在这样的关系中进行着两者之间的比较，盆栽占55%，而友谊则是45%，两者几乎持平。

我希望你们能喜欢我的演讲并启发大家建立友谊。自己一个人制作盆栽固然很好，但如果可以与伙伴们一起进行创作则会更加有趣。

丹麦盆景欣赏
Denmark Penjing appreciation

The Bridge of Friendship
— Bonsai

Speaker: [Denmark] Tommy Niesen   Reorganizer: CP

Hello everybody!

Some might know me and many does not know me so I will try to introduce myself.

My name is Tommy Nielsen and I am the president at the Danish Bonsai society. I was elected 3 years ago as president and is the youngest president the society ever had. I am 31 years old and is married to Monica my lovely wife, we live in a small town in Denmark called Daugbjerg. Daugbjerg is very small and there is only 600 people living here, here are no sounds as in the bigger cities, the most common sounds is from birds, foxes and deer.

The Danish Bonsai board is a very young society only 32 years. Every year we have a festival where we find the best Bonsai and Shohin composition, this event is over a weekend in April or may. I look forward to this weekend every year, even though I do not have any thing that is ready for the festival.

Some of the things I like very much about the festival is that you meet old friends but also make some new, you get inspired by other peoples work and you can admire their trees.

When the society was born the society had nearly 700 members, people were meeting in liberty with their materials and were helping each other. Over the years the society got bigger and bigger and in the early 90s there were 780 people making Bonsai in Denmark. After that there became lesser and lesser people in the society and today we are a small society with only 253 members left.

What went wrong?

Many of the members are older people and many of the older members back in the 90s have passed away, Bonsai became an opportunity besides all the other activities that you can go for.Younger people do not see what is so fantastic about Bonsai and do not understand why this is such a nice hobby and it is hard to tell them when they do not have the time for listening. One of the other problems is that the trees which were young ones is now very powerful Bonsai where you can see that there is many hours of work, maybe some gets scared about this.

When I was asked to make this speech and talk about Danish Bonsai culture one thing came on my mind. Why do you make Bonsai?

When I saw a Bonsai the first time I was only 9 years old, it was in a Japanese garden in Denmark. My fascination of Bonsai and Asian culture started when I was 7 years old I began exercise karate, I hated soccer and other team sports and I was teased in school so the only place I could go was to the dojo and exercise. My parents were given a Bonsai when I was 10 years old, as I remember it not a very powerful tree but it fascinated me very much. The tree died as we did not know that it should be repotted and all the old soil should have been changes. Unfortunately the plant were from a supermarket and there were no guidance about this so the tree died. Time went on and as I was turning 18 years old, my parents gave me a book called Bonsai for beginners and that was where it all started. The book was not very good but it taught me some of the simple things about Bonsai. I found myself a local club and got started from some of the older

丹麦盆景欣赏
Denmark Penjing appreciation

丹麦盆景欣赏
Denmark Penjing appreciation

The Annual Forum 国际年度论坛

丹麦盆景欣赏
Denmark Penjing appreciation

guys. In the beginning I saw a lot of movies and one of my new heroes where Kevin Wilson and his great work with power tools. I started out finding big materials and began to work on these, it needed a lot of work and I found myself relaxing and enjoying it. I used it as a way to stress down and find my inner balance, this was something I could do when I wanted to do it, I did not need to fit in to other guys plans and I could do it for hours and still come back a few days later to start over again on a new material.

The first 7 years went by and I tried to look at how the Japanese guys made their materials and tried to do the same. Then my hobby took a new way, I was introduced in to Shohin Bonsai by Morten Albaek, he showed me the secrets hidden in the small trees and the great power they were having. Morten inspired me to do Shohin Bonsai and Torben Brendfeldt and Yvone Garbaek told me how to keep these trees. They also told me how to display the trees in a fair and they made me thinking about what Bonsai really is.

In 2012 I meet Vaclav Novak as he was hired to make demos and workshops at our yearly festival that year. As Vaclav also were judging the Bonsai I needed to know which trees that he picked for the different awards and why. Vaclav told me how to do Bonsai in another way then the Japanese do it, he told me that Bonsai needed to be natural and they should tell us a story from the nature even though they were not yamadori. Bonsai is all about creation and fantasy, it is like painting a picture of writing a story book, they shall tell a story about the guy or girl who is standing behind it and it was all new to me but made me look at Bonsai in a different way.

2013 I meet one of my idols Kevin Wilson and his power tool, this was very big to me and Kevin was all you could ask from someone like him. Kevin told many funny stories and we became friends, he also said that he did not make the judgment at the festival so we had to do it ourselves. The reason why Kevin did not want to do this was because he believed that all Bonsai were equal and it was the owner's job to tell a story through the trees. Kevin actually said the same thing as Vaclav, but he also said one more thing he said people should be making Bonsai out of nature, and look at the trees in their own country how they look, and use them to get inspired in how to do Bonsai.

Over the last 2 years I have been thinking about why I make Bonsai, not that I was considering stopping but I have asked myself why you are making Bonsai. You can try and ask yourself why I make Bonsai. For me the answer was not easy to find, I had to ask myself many times over and over again. I ended up with two answers.

(1) By making Bonsai I can express myself and my work through my Bonsai, I can decide to cut a branch, repot in a new pot, wire and restyle if this is what I want. Bonsai also help me finding inner peace and helps me to relax and forgetting all the things that is stressing me. But the most important thing when we ask ourselves why we make Bonsai is that it gathers people and creating friendships across borders all over the world.

(2) Bonsai get friendships to grow despite the age differences between young and old. This is something unlike, how many hobbies can you speak about where a 15-20 year young guy/girl is having a real good friendship with a man or a woman wish could be their grandparents. If I shall tell how the 2 things are compared to each other I wanted to say that Bonsai is 55% and friendships are 45% so almost equal.

I hope you all enjoyed my speech and have been inspired to make a lot of friendships with each other. It is ok to sit by yourselves and do Bonsai but it simply is more fun if you have company.

丹麦盆景欣赏
Denmark Penjing appreciation

2013 唐苑的世界盆景对话
DIALOGUE TO THE WORLD PENJING

石头的传说
Tales of stone

发言人：[意大利]玛利亚·基亚拉·帕德里齐　图文整理：CP　　Speaker: [Itsly] Maria Chiara Padrini　Reorganizer: CP

玛利亚·基亚拉·帕德里齐 国际盆栽俱乐部意大利理事
Maria Chiara Padrini Italian Director of Bonsai Clubs International

意大利观赏石
View Stone of Italy

　　我想我的发言与盆景的话题无关因为我感觉大家都在谈论盆景和盆栽的话题，但我是一位赏石爱好者，所以我将讲述赏石的价值。

　　从2005年至今我已经多次来中国旅行。但仍然无法让我充分领略这个国家的山河美景以及充分了解对人类历史带来深远影响的博大精深的中国文化。

　　在此，我想感谢中国盆景艺术家协会苏放会长能给我这个机会再次来到中国并加深我对这方面的了解。我也十分赞赏他们对收集来自世界各地的不同文化、经验和知识所做出的努力。就盆景、盆栽和赏石、水石艺术爱好者而言，赏玩的主要目的之一在于可以开创新的联系并加强各国盆景与赏石爱好者间的友谊。这次论坛的最终目标就在于此。

　　作为赏石爱好者，我希望介绍石头给大家，让大家知道在意大利我们如何对石头产生浓厚的兴趣。我们国家赏石这项活动开始于40年前，所以我们还是初学者，但在我们国家到处都能发现奇特怪异、带给人灵感的石头，所以这促使爱好者收藏并热爱赏石文化。

　　随后，我想通过幻灯片告诉大家一些与这个工作相关的

动物系列——狗
Animal series-dog

动物系列——兔子
Animal series- rabbit

动物系列——鸟
Animal series-bird

The Annual Forum 国际年度论坛

动物系列——海狮
Animal series-sea lions

动物系列——北极熊
Animal series-polar bear

动物系列——大象
Animal series- elephants

事情,这个展示与其他的幻灯片展示有点不同。即我们想试图通过石头来传播一些建议并通过赏石激励人们某些方面的能力。

我们对石头的热爱体现在对石头的找寻、挑选、清洗和雕刻的各个阶段,但它们对于我而言意味着"赏石"。

寻找到每个人心仪的石头需要我们投入大量的精力,例如:从精神层面、心理上和个人情感方面,才能找到属于每个人的石头。

当这条道路不再受到个人因素的影响时,石头则向我们介绍着巨大且客观的艺术空间。因此它们成为了艺术品、文化遗产以及与大家沟通的一种方式。

也许"石头的传说"是一个有着雄心壮志的提案,即使它无法进行分类也没有进行过隆重的展示,但是它却试图通过展示石头最原始的一面来唤醒它自身的暗示能力。

我也曾经在各种可能的情况下强调自然界人类存在的意义,并证明我们与石头之间的联系。纯朴的、令人怜惜又普通的人类正如质朴美丽的石头一样,这些都是上帝赐予人类最美好的礼物是普遍存在的,并且每个人都拥有它。

人类因为常常忽视它,所以无法发现那令人难以置信的蕴藏在石头内部的美。这需要善于观察的双眼和充满激情的心才能发现它们。

人物石头图片欣赏
Figure Stone Picture

Tales of Stone

Speaker: [Itsly] Maria Chiara Padrini Reorganizer: CP

I think that my speech is out of the topic of Penjing because I feel your speeches are about traditional Penjing or bonsai, but I'm a stone lover, so I'll speak about stones' value.

I have made many trips to China, the first time was in 2005, but they would never be enough to know the beauties of your country and the vast culture that has influenced the history of humankind.

I am therefore grateful to CPAA and Mr. Su Fang for offering me this opportunity to come back again and deepen my little knowledge. I also appreciate the effort of bringing together different cultures, experiences and knowledge of many countries of the world. One of the main goals of the enthusiasts in the arts of Penjing, Bonsai and Viewing Stone or Suiseki is to open new connections and foster friendship among individuals and peoples. Discard the seeds of brotherhood is more important to me than to educate a tree or admire a stone. It should represent the ultimate goal of our work.

As a stone lover, I wish to introduce stones to you to know how in Italy we're interested in stones. This started from almost 40 years ago in our country, so we're beginners, but we can find romantic yield somewhere, so this makes enthusiasts to collect and love viewing stones.

Subsequently, I'd like to introduce some things of this job that thought of trying to transmit the suggestions and the ability to excite in a stone through presentations which are a little different from the usual presentations.

Looking for stones, picking up, cleaning, carving on it, are phases that accompany our being fond of stones, but they represent a preamble to what, according to me, means "stone appreciation."

The mental, psychological, emotional aspect let's find in a stone so many inputs to cross more intimate and personal ways.

When this road runs free from a personal submission, the stone introduces us in well vast and objective artistic spaces. It becomes therefore work of art, heritage and a way of communication for everybody.

Perhaps "Tales of stone" is an ambitious proposal thought out of every classification and rigorous or emphatic presentation, however of trying to arouse the suggestive abilities from the stone shown with no make-up than its clean face.

I also tried, when possible, to underline the man's presence, to make evident the bond which ties us to the stone. A simple, poor, common humankind as the stone is in the gift of its ubiquity and access for everybody.

Human beings often neglected, but not deprived of an unbelievable beauty, hidden in the folds of the life as stones asking for a careful eye and an impassioned heart to let them to be found.

山,总是女王
她的姿态唤醒我们内心的形象,
告知我们细节、色彩、声音、季节的演变
The mountain - the queen is always her
Her shape awakes the images we keep inside
Overall let us evoke details, colors, sounds, seasons…

云彩环绕着她
The clouds surrounding her

有时几乎完全遮挡了她
Sometimes mists almost hide her

有了石头,你也可以开始一段故事
And now that you have a stone, you can start a story…

"和谐" 相思 Celtis sinensis 吴成发藏品
（见《中国盆景赏石·2012-3》折页）

"Harmony". Chinese Nettletree. Collector: Ng Shingfat
(Folding Page of the *China Penjing & Scholar's Rocks* 2012-3)

文：徐昊 Author: Xu Hao

凡高手作盆景，必有用意，或以情入景，或因树造境，是谓立意。

该作一本双干，以基部相连，连接也颇动感，双干态势妖娆，出枝如展臂，双双敞开怀抱，作相拥未拥之态。

两老者静坐其右，看似闲聊或是向对方讲述这里曾经发生的故事，讲者认真，听者入神。

作品以拟人化的手法，将相悦者相遇时美妙瞬间以盆景的形式定格下来，景中描绘的是人？是树？万物皆有灵性，览者各自理解。

作品线条清疏飘逸，整体气脉灵动，有情有景，情景交融，是一件可赏性颇佳的作品。

Any Penjing master starts to create, he must have his intention. Either he puts emotion into the scenery or he creates a scene based on the tree. That is to build the conception.

This work has one main stem with two trunks connected by the base, which creates a dynamic picture. The posture of two trunks is enchanting and those branches pushing out like opening arms are ready to hug but have not embraced yet.

Two old men sitting on the right side of the tree, it seems that they are chatting or telling each other about what happened here. The teller speaks carefully and the listener is enthralled.

The work with the personification seize the wonderful moment in the form of Penjing when pleasing people meet each other. What is depicted in the picture, a person or a tree? Everything has spirit and the viewers have different understanding.

The work's line is fresh and elegant. The whole idea is in sequence with passion and view, and a fusion of feelings with the natural setting. It is an appreciating good work.

"和谐" 相思 *Celtis sinensis* 高120cm 吴成发藏品 苏放摄影
"Harmony". Chinese Nettletree. Height: 120cm.
Collector: Ng Shingfat, Photographer: Su Fang

榆树 Ulmus pumila 黄经洲藏品
（见《中国盆景赏石·2012-8》第22页）
Elm. Collector: Huang Jingzhou
(Page 22 of the *China Penjing & Scholar's Rocks* 2012-8)
文：李新 Author: Li Xin

引发这篇随感的是《中国盆景赏石·2012-8》中的一盆榆树盆景，作者黄经洲，从风格上看，应是台湾地区的作品，典型的密枝作风，温润工整，精巧细腻，枝法造型、管理养护均极精深，重要的是，有难得一见的自然气息。

在陈述优点之前，先说一下此树的缺憾。其一，在整体构成上，树冠繁茂而基干单薄，有不胜重负之感，尤其根盘至第一转折处，过于短促，与其上部比例悬殊，且青嫩有余，苍老浑厚不足，使该作的艺术魅力大打折扣。其二，未能很好地控制顶端生长优势，使得树体上部尤其是居中的枝条稍粗，一方面加重了上部比重，另一方面与左下垂枝粗细对比明显，此垂枝恰是全树的精华之一，使得本应强化的枝条因此被削弱。其三，盆钵使用有待商榷。此桩配盆较单薄狭小，与丰满宽大的树形不相协调，若取一件体量与之相匹的椭圆抑或圆形盆钵，扩大土壤面积，增加底部分量，似乎更为得体，也更能承载得起如此丰硕的树冠，从而削弱头重脚轻之弊。

尽管如此，此作所呈现的典型的树木风貌、弥漫的自然风味还是打动了我，让我驻目留连。就枝法与树冠风姿而言，可说是师法自然、成功复制大树风貌的一个典型范本。

这里，不见寻常盆景的制作格式与规范，没有做作的弯曲，亦没有匠气的堆砌，而是遵照自然树木的生长规律和典型风貌进行归纳塑造，同时也是一次精准地复制与还原。作品右下数根枝条，繁茂驳杂却不臃肿闭塞，它们相互依托相互簇拥，安然自在地伸展倾垂，像从自然中截裁过来一般，不见丝毫匠痕；左下垂枝，尽管分量较弱，不甚抢眼，但细细赏来，跌宕流泻，宛转绰约，仿若戏曲舞台上佳人回眸转身抛出的水袖，轻盈而曼妙。不禁感叹：若无深入自然详尽观察之功，绝无眼前这一派天然气象，抑或，作者天赋使然。

这就与充斥我们视野的绝大多数盆景拉开了距离。

我曾在一篇文章中说过："即便我们常说的'缩龙成寸'，即寻常大树形态的缩小和复制，在实际落实中也不那么常见。常见的是攀扎细致、分布均匀、层层叠叠、貌似大树的规整植物。说貌似，是因为这些盆栽看似一株大树，然细加打量，则与大树的自然形态相距甚远，布局生硬呆板、缺乏有机联系、千篇一律、人云亦云者比比皆是……有时让人赞许的反倒是作者的耐心和'制作工艺'，若说'自然'，则未必。"

这样的作品，无论展评会还是专业刊物，十之有九。在我看来，它们的"盆景味儿"太浓，与树木形态差距较大，也就远离了自然。

这里出现一个悖论：盆中的一切，均来自大自然，是大自然的产物，其初衷也是为了表现大自然。然而，经过精心设计与摆弄，最终却以"不自然"的姿态出现，让人不由蹙眉。

由此引发一个问题：脱离了大自然的盆景创作，其意义和价值何在？

有时会想到手工艺品，而且是品相不佳的手工艺品，因为真正好的工艺品足登大雅，亦可承载时代精神。莫说商周时期的青铜器皿，厚重恢弘，无与伦比，即便宋元瓷器、明清织锦，工艺水准和艺术表现也都精妙绝伦，代表了那个时代的风范。

与之相形，我们的"制作工艺"即刻见拙。尽管，商业价值与艺术价值不是一个概念，更不能画上等号，但有时候，把它拿出来衡量一下我们的盆景作品，也不失为一个小小参照。以上提及的工艺品，它们中任何一个与盆景相比较，商业的天平将会出现怎样地剧烈倾斜，不是一目了然么？

商业价值，在我眼里从来不是考量标准，但以目前社会对盆景的认可度和市场给予它的定位看，我以为大致公允。

不要说别人看不懂、不理解，一个有着5000年文明的泱泱大国，怎么会缺少锐利敏感的目光呢，问题是你能不能让这样的目光被触动。还是要从自身找问题，这是一个理性清醒的人应持的基本思路。

点评 Comments

榆树 *Ulmus pumila* 黄经洲藏品 苏放摄影
Elm. Collector: Huang Jingzhou, Photographer: Su Fang

点评 Comments

榆树 *Ulmus pumila* 黄经洲藏品 苏放摄影
Elm. Collector: Huang Jingzhou, Photographer: Su Fang

盆景目前仍然是在圈内玩赏、流转，远未进入社会与公众的视野，更未进入真正的艺术品市场，不被主流艺术承认已是不争的事实。若说原因也简单，除了日常保养较为不便、不利流通外，关键是文化、艺术含量较低，不能引发明眼人的兴趣。

如何走出窘境呢？或者说如何提高自身的艺术价值？

我以为首要的任务，是要让我们的盆景先变得自然起来，在与大自然亲密接触、融合的基础上，再谈其他。

其实"自然"，或者说要求这些盆中之木像自然中的一棵树，只是艺术创作中一个最低最起码的要求。就像画家描绘模特一样，"像"只是基础，如果连像都做不到，谈何表现人物表情和精神呢？再进一步讲，如果无法如实描绘对象，又如何准确传达更加复杂、丰富的思想情感呢？

然而我们的盆景现状，就像一群面对静物写生的学生，因为画不像而被长期滞留在低年级画室，无法进一步深造。也因此，从最基础的功课做起，认真向大自然学习，培养准确塑造大树的能力，是摆在大多数盆景人面前的一个紧迫问题。

为避免误解，在那篇文章里，我说："这里，非是以树木的自然形态为旨归，而是希望能在认真观察、准确掌握的基础上，汲取英华，抛却凡常，创作出既能反映自然微妙变化，又具个人情怀的盆景作品。"

如若不能，退求其次——"其实准确模拟，也非易事。自然界中那些优美奇宕的树相（尤其某些转折奇诡、精妙难言的枝干变化穿插），足以让我们心旌神摇，心摹手追的了，若将其准确还原到盆中，也将是了不起的贡献。"

之所以说"了不起"，是因为在我们的视野里，这样的作品少之又少，这也是我为什么要以这棵榆树作引子来谈的原因。

坦率地说，这件作品，是我们前行路上的一个起跑线，或者说一个基准点。它的风貌尽管自然，但是个性还不鲜明，只是一棵较为写实的树而已，再苛刻一点讲，局部精彩，整体平淡。

而真正有意思的创作，应该是在这样的基础上再进一步，也就是以上所说：汲取英华，抛却凡常，创作出既能反映自然微妙变化，又颇具个人情怀和形式意味的盆景作品。

只有这样，我们的盆景才能真正进入艺术创作的层面，并在造型艺术规律的统摄下，以盆钵、木石为载体，以个人识见、志趣为主导，充分吸收融合姊妹艺术养分，发展探索自己的表达方式和语言，并不断深化延伸，穷尽其可能，进而呈现多姿多彩、个性林立的创作局面，迈向更高的艺术殿堂。

说来痛快，落到纸上也就几行，真正做到，何其难！

然而舍此，有第二条路可走么？

算是一种想望吧。

This essay originated from an Elm Penjing, which was published on the eighth issue of 2012 China Penjing & Scholar's Rocks. The creator is Huang Jingzhou. The Penjing is supposed to be the work from Taiwan area according to the style of typical thick branches, being mild, neat and orderly, delicate and exquisite, and the fine and exquisite branch modeling and management and maintenance. The most important point is the rarely natural flavor.

Before stating its advantages, I will talk about its shortcomings. First, on the aspect of the overall structure, the crown is lush but the trunk is thin and weak, which makes the sense of overburden. Especially, the part from root to the first turning point is over short, having wide gap with the proportion of its upper part, and the tree is too young to show its age and vigorousness, so that the artistic charm reduces greatly. Second, apical growth advantage is not controlled well, so that the upper part of the tree is thicker, especially branches in the middle. On the one hand, it increases the proportion of the upper part. On the other hand, its thickness is obvious when contrasting with the left drooping branch, which is one of the essences of the whole tree, weakening the branch that should have been strengthened. Third, the pot using shall be considered further. The pot for this tree is small and narrow, which is not coordinated with the plump and the large tree shape. It seems to be better to take the oval or round pot with which matched size to expand the soil area and increase the measure of the bottom, which can carry such bountiful crown better, so as to weaken the disadvantage of top-heavy.

Nevertheless, I am touched by its typical style and features of the tree and the diffusing natural flavor, which causes me to stay and appreciate. On the aspect of branch and the charm of the crown, it can be called as a typical example of naturally and successfully copying the style of tree.

It has no usual making format and specification of Penjing, no artificial bending, nor unimaginative concoction. It models in compliance with the growth rhythm and typical style of natural trees, which is also a precise copy and reduction. Some branches under the right side are lush but not crowded. They rely on each other, surround each other, stretch and droop naturally and freely, just like that they are cut from the nature without any artificial sign; though the left drooping branches take a small part and are not attractive, they are flickering and graceful when appreciating carefully, just like light and delicate long sleeves thrown out by beautiful women on the stage of Chinese opera when they turning around. I have to say if there is no detailed observation going into the natural deeply, such a great natural scene can't be seen at the moment—or it is made by the natural gift of maker's.

Therefore, this work is superior to most Penjing works in our vision.

I once said in an article: "Even as we often talked 'shrink the tree to an inch' which is the down size and reproduce of usual trees model, is not common in the practical implementation. It's common to see regular plants like big trees which are tied carefully, distributed evenly. I said 'like' because these Bonsai look like big trees, but look at carefully, the tree's natural shape are wide apart, they are stiff in layout, lack of organic links, follow the same pattern and they can be found everywhere. Sometimes, the patience and 'craftsmanship' of the authors' should be praised. As for 'natural', it may not.

Most works displayed in the exhibition or published on the professional journals are similar to these kinds of works. In my opinion, with strong "Penjing style", they are far away from normal trees shape and nature.

Here is a paradox: all the plants in the pot are from nature and they are the products of nature, its purpose is to show the nature. However, after careful designing, they eventually appear in an "unnatural" shape which makes people can't help frowning.

This leads to the question: what is the meaning and value of the Penjing creation out of nature?

Sometimes this reminds me of handcrafts, especially handcrafts with poor quality, because the really good crafts are in good taste and can carry the spirit of the times. Don't say the bronze vessels of Shang and Zhou dynasties, which are thick, majestic and incomparable, even Song and Yuan Dynasties' porcelain, Ming and Qing Dynasties' brocade which are absolutely exquisite in the standard of craftsmanship and artistic expression, they can represent the demeanor of then era.

Compared with them, our "craftsmanship" becomes inferior to be seen instantly. Although commercial value and artistic value is not one concept and not equal to each other, sometimes it is a little reference to measure our Bonsai works. Take any one of the above mentioned crafts to compare with Penjing, is it clear how the balance of business will be tilted sharply, isn't it?

In my eyes, commercial value is never the standard to measure Penjing, but I think it is basically fair from the point of social acceptance and market positioning of Penjing.

Do not blame others that they can't understand our works. How a great country with 5,000 years of civilization history, how could we lack of sharp and sensitive eyes, the problem is that whether you can shock such eyes.

We should find the problem in ourselves, and this is the basic idea that a rational and lucid people should hold.

Currently, far from entering into the view field of the social and public, let alone the real art market, Penjing is still appreciated and circulated in our circle. It is an indisputable fact that it's not recognized by the mainstream art. The reason is simple, in addition to the routine maintenance is not convenient to circulate, the key reason is that it can't attract knowledgeable people because of its low cultural and artistic content.

But how do we walk out of the dilemma? Or how do we increase the artistic value of Penjing?

First of all, Penjing should be natural increasingly, and other aspects should be followed based on its close connection and integration with nature.

"Being natural", or trees in these pots should be similar to a tree in nature, and this is the minimum requirement for artistic creation. Just like an artist painting model, "similar" should be the base, if we can't do a Penjing looks like a tree in nature, how shall we talk about facial expression and spirit? Furthermore, how could the thoughts and feelings which are more complicated and more abundant are expressed precisely if the object was not presented faithfully?

However, the current situation of our Penjing industry is just like a group of students sketching in the face of still life, but they are retained in low-grade classroom from further study due to their poor sketching. Therefore, starting from the most basic to learn from nature and cultivate the capacity to shape large trees becomes an urgent problem to be solved by most people involved in Penjing industry.

In order to avoid misunderstanding, I wrote in that paper that "the natural shape of a tree is not the final purpose here. We hope to create a Penjing representing the subtle changes of nature and personal feelings by absorbing the essences while eliminating the ordinary based on careful observation and accurate grasp."

If failed, then seek for the next best — "it is not easy to imitate accurately. Those wonderful shapes of trees (especially some branches with rare twist and extraordinary changes) fascinate us so much that we all are eager to imitate them by ourselves. It will be an outstanding achievement if they are represented within pots."

I said "outstanding" because such works are few and far between in our visual field, and that's why I took this elm for opening words.

Frankly speaking, this work is just a starting line, or a benchmark on our road. Despite its style and feature looks like nature, its personality is not so distinctive, and it is just a relatively natural tree, or more strictly, it is brilliant partly but insipid wholly.

Truly outstanding creations should go further based on this, as we have mentioned above: to create Penjing not only to reflect the subtle changes of nature and rather personal feelings and form implications by absorbing the essences while eliminating the ordinary.

Only by this, our Penjing can enter into artistic creation level and under the control of the rule of formative arts by taking pots, wood and stone as carrier, personal knowledge, interests as the leading, absorbing and integrating with nutrients from sister arts, developing and exploring our own expression and language, and extending constantly as far as possible to present a colorful creation situation with various personalities.

It is easy to write a few lines on the paper, but it is never easy to achieve.

Is there any way to achieve other than this?

I hope that we can make it.

树木移植的国际做法（连载二）
The International Method of Tree Transplantation (Serial II)

文：欧永森 Author: Sammy Au

作者简介
欧永森，中华树艺师学会会长，香港树木学会会长，国际攀树学会会员，国际棕榈学会会员，英国皇家园艺学会会员，香港园艺学会终生会员。

　　讲到这里，终于可以开始讲移树的步骤了。对一些行外人来说，如果前面的一大堆不讲，后面的再讲也没有意义，因为科学道理不先讲清楚，谁会相信您呢？

　　正当移树的步骤，大概可以分成：

1. 移前准备
2. 包扎树冠
3. 起挖土球
4. 包装土球
5. 吊运上车
6. 中途运输
7. 新树洞的事前准备
8. 树木卸装和工地内运
9. 树木种植、现场修剪和实时护养
10. 支撑、覆盖物、施肥和长期护养
11. 定时检查

　　以上各点，缺一不可。成功地移树有如做大手术一样，不是随便勾起、挖洞和吊上去，然后浇浇水那么简单，也就说明了为何西方国家的种树与我们的有大不同的地方，不是谁比较聪明，而是谁比较认真。

　　城市里面大部分的绿化树木，都是移植来的，如果这一关把不好，只能够是"周瑜妙计安天下，赔了夫人又折兵"吧！

柯家花园仿古石盆 系列欣赏

The Appreciation of the Ke Chengkun's Antique Pot Series

欣赏网址：http://www.xmkjhy.com
欣赏咨询电话：18650163765

301# 仿古石盆 外径2m 内径1.56m 高1.15m 内高0.8m
树种：真柏 高 1.16m 宽 2.08m 直径 0.4m 柯成昆藏品，柯博达摄影
301# Imitation Ancient Rock Pot. Outside Diameters: 2m, Inside Diameter: 1.56m,
Height: 1.15m, Inside Height: 0.8m. Chinese Juniper.
Height: 1.16m, Width: 2.08m, Diameter: 0.4m.
Collector: Ke Chengkun. Photographer: Ke Boda

Conservation and Management 养护与管理

1. 移前准备

这里也分成几个方面:

①如果是行道树,其质量必须符合某些国际标准,否则只会徒劳。苗量差的可能会不停发病也站不稳的,慎用。

②如果是其他树(公园、小区、医院等),其质量也必须符合某些既定的国际标准。

③如果是老树大树(从不建议移植),在评议过后还是决定要移的话,若没有质量问题可言,只能接受现状。

无论是什么树,长在什么地方(苗场、山上、城市里面等),都有共通的移前准备步骤,包括:

①移前的1~3个月,给足水肥,让根叶蓬勃发展。在这里,施肥要小心,树木需肥量极少,看山上长的树就知道。

②条件许可的话,可以进行"断根再生",促进吸收根的发生。"断根"并非随便把根切掉,然后放一些培植土打包,让新根再长;是要懂得挑选切哪一条、切多少和留多少。乱切根,日后会生长不良或在风中倒塌。

③把树冠的所有的枯、死、病、断、交叉枝、弱势枝等统统切掉,即树木修剪学里面的"树冠清理"(Crown cleaning),以防日后发生病虫害。这里要学习怎样去找这类枝条,和用什么技术去修。

④挂上标号,以防日后工人找不到或找错。

2. 包扎树冠

在进行树冠包扎之前,有一点是非常重要的,就是先进行修剪。把没有必要的枝条也运到工地去,只会增大包装的难度和运输的成本,运到工地以后又要把它剪掉扔掉,费时费事。

在包扎前修剪之时,又有两种做法:

①如果是小苗(指1~4年生的树苗),应该要进行最后一次的"结构性修剪"(Structural pruning),来使树冠形状变小,保证结构性的安全。

②如果是大苗、大树或老树,则进行最后一次的"树冠清理",来减轻其大小和重量。要留意,这里是要尽量剪掉死去的枝条,而不是剪掉活着的枝条,活枝只在迫不得已的情况下才剪掉,因为它里面储藏了养分,如果还未转移到树体就去掉,太浪费了。

顶芽要尽量不剪,特别是中央主干的顶芽。如果剪了,就把制造"生长素"的部位一举端掉,树体会马上发动"水芽",然后长得乱七八糟。如果把中央主干的顶芽也剪掉,日后的树干只会长宽不长高,这在盆景制作中经常用到。

树枝的包装,也要非常小心,如果最后会被包装材料压断的话,那倒不如在先前剪掉,起码切口会比较平整,易于"创伤板隔化"(Compartmentalization of Decay in tree)的修复行动进行。

树冠包装的形状,要依照树冠原来的形状稍微压缩一点,在运输过程中,一般又以尖锥型或圆型最抗风。拉的时候,一般直径2cm以下的枝条弯曲的多一点,粗大的很容易被拉裂,到最后也要被剪掉。

绑拉枝条时,最好是从中央主干两边的枝条开始,一层一层地往中央方向靠拢,最后变成圆锥形或圆形。一般大卡车的货箱宽度只有2.5m,如果包装好的树冠直径超过这个的话,可能要在运输中做出特别安排,这也是为何西方国家一般不搬移大树老树的一个原因。

包装材料方面,在华南的高温、多湿、强日照地区,以两层的遮光网来打包,最为理想。一则便宜,二则透风透光,不会把树冠"蒸"熟,三则可以在一定程度上挡风,可耐几小时的运输。

有人会问,那我喷"抗蒸发剂"(Anti-transpiring)到树冠上面,那就不是可以防止风干了吗?研究指出,这些油蜡性的喷剂,能够堵住蒸发气孔,有如跑步10km而不出汗一样,效果可能会差强人意。所以,除了在个别合适气候条件和树种情况以外,不建议使用。

3. 起挖土球

绝对不能用挖土机械来起挖土球(特制的"搬树插"Tree spade 例外)!

挖土机的挖斗是用来挖土挖石头的,不是用来断根的。那个挖斗一挖进去,可能碰到一整片的根系,都给它撕裂弄翻,以后再作修补整理是一件非常困难的苦差,而且没有保证一定能够修复。所以不管是小树、大树或老树,用人手切挖是国际上公认的最理想做法。

土球的直径一般要在胸径的10倍以上,或一直挖到树冠滴水线(Drip line)以下,才能保住重要根系。换句话说,一棵1m胸径的大树,就要挖10m直径以上的土球,这个连同树体一起,重量可能已经超过100t,起吊和运输都有一定困难,这也是西方国家一般不喜欢搬大树老树的原因之一。

断根的手工具必须锋利。因为根系的修补和树体一样,都是靠"创伤板隔化"(CODIT)的顺利进行,撕裂的根系很难进行"创伤板隔化"。不管日后喷多少肥料、杀菌剂、生根素等,对根系的修补实是无济于事,研究也不支持它们的使用,因为没有多少效果。

土球的形状不是绝对固定的,大部分是圆形的,也有方形的,因为在同一长度里,方形带的土和根系比圆形的多。有些时候也要看固定根的走势来决定土球的形状,固定根越长,那边的土球也越长,没有绝对限制。

土球的深度一般在1m之内,再挖深了可能只是搬土,而不是搬根了。

【未完待续】

"塑韵" 百色石 长14cm 宽9cm 高17cm 李正银藏品 苏放摄影
"Sculpture with Special Appeal". Baise Stone. Length: 14cm, Width: 9cm, Height: 17cm.
Collector: Li Zhengyin, Photographer: Su Fang

CHINA SCHOLAR'S ROCKS
赏石中国

本年度本栏目协办人：李正银，魏积泉

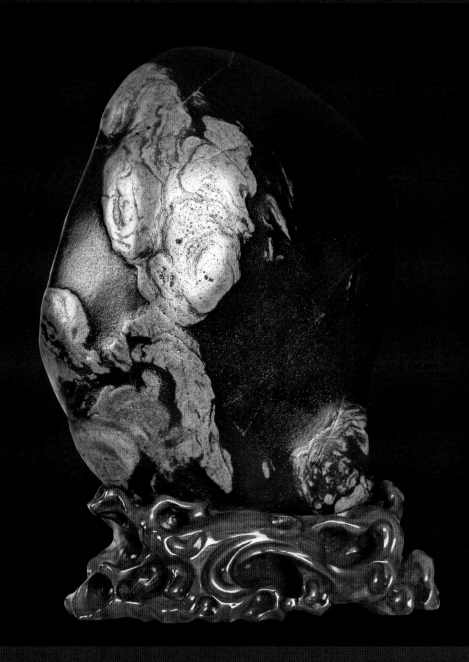

"一代天骄" 广西三江石 长20cm 宽7cm 高16cm 魏积泉藏品
"God's Favored One". Guangxi Sanjiang Stone. Length: 20cm, Width: 7cm, Height: 16cm. Collector: Wei Jiquan

中国古今名石简谱（连载八）

Chinese Famous Rocks Notation (Serial VIII)

文：文甡 Author: Wen Shen

中国的印石和砚石，始用于先秦。"文房"即书房的概念，始于南唐后主李煜。中国文人砚石的滥觞始于唐武德年间，兴于两宋。南唐文房文化，与北宋多有契合之处。中国文人印石始于元代翰林大学士赵孟頫，兴于明代两京国子监博士文彭。中国文房名石，即包括文房名印石和文房名砚石。

中国文房名砚石与名印石，与其他大量的砚石与印石，有着本质的不同。笔者将在本谱中，与您一起走进中国文房名石的家乡，饱览那里秀美的风光、经历寻访奇石的艰险、体会文房美石的特质与风韵。

笔者造访青田石大王倪东方先生的陈列室"惜石斋"

一、中国文房名印石

印章作为凭信之物，始于先秦。学者罗福颐编著的《古玺汇编》，收录了3800枚战国姓名印，可以看出先秦印鉴的普遍使用，但大多只是私印，以铜为质。

汉印艺为宗

篆刻艺术家奚冈曾于印边款识："印之宗汉，为诗之宗唐，字之宗晋。"汉印在中国印学史上，是一座不可迄及的高峰，它的大气、古拙，两千年来为印界仰止。这与汉代官员的文化素养有直接关系。汉《说文解字·序》说："十七以上，始试讽籀书九千字，乃得为吏。又以八体试之，郡移太史，并课最者，以为尚书史，书或不正，辄举劾之。"认字多、书法好，才能作秘书官，这为汉印制作打下基础。治印艺术，讲究书法、章法、刀法三法俱佳，相辅相成。明甘旸《印章集说·章法》说："布置成文曰章法。欲臻其妙，务准绳古印……那让取巧，当本乎正。使相依顾而有情，一气贯串而不悖，始尽其善。"汉印章法，整体均衡协调、朱白兼顾、气息贯通，技艺高深。汉印的刀法也十分精湛，不仅印面线条流畅，印底也光洁工整，突显功力。

魏晋唐宋印艺不足取

中国的印章篆艺从魏晋开始走下坡，至唐宋至谷底。但收藏鉴赏印的使用，却是对印学的极大推动。唐张彦远《历代名画记》中指出："明跋尾印记，乃书画之本业耳。"把印章艺术提高到与书画同等的地位。甘旸《印章集说》："上古收藏书画，原无印记，始于唐宋，近代好事者耳。"五代十国南唐后主李煜的书画，被宋代皇帝掠去，书画上大都钤有"建业文房之印"。金人缴获宋徽宗、钦宗两帝玉宝28方，各种收藏、吉语等印35枚，可见当时书画钤印已经普遍。宋米芾在《书史》中说："大印粗文，若施于书画，占纸素字画多，有损于书帖。王诜见余家印记与唐印相似，尽更换了，作细圈，仍皆求余作篆。如填篆自有法，近世填篆皆无法。"米芾指出了宋印滥用的弊病。苏轼《东坡尺牍》中有给米芾的信："卧阅四印奇古，失病所在，……印却纳。"从信中可以看出，米芾也为东坡治印，两人对印之积弊，皆有同感。

元明清文人印学兴起

元代书画家赵孟頫在《印史》中批评不合古意的流俗，提出复兴汉印古雅、质朴的风尚。元末刘绩《霏雪录》记："以花药石刻印者，自山农始也。山

China Scholar's Rocks 赏石中国

"快乐的小精灵" 戈壁玛瑙 长30cm 宽27cm 高39cm 魏积泉藏品
"Happy Elf". Gobi Agate Stone. Length: 30cm, Width: 27cm, Height: 39cm. Collector: Wei Jiquan

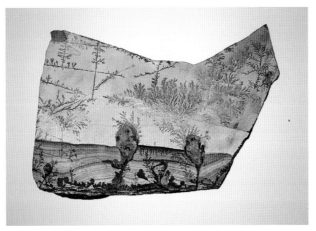

图15 "山雨欲来" 八公山模树石 胡芝兰藏品　图16 "楼兰古色" 八公山模树石 胡芝兰藏品

地下溶蚀，其龟纹形状有似乌云翻滚的卷云纹，有一道道似斧劈至底的斧劈纹，也有似层层树叶堆积的千层纹、还有豆瓣纹、莲花纹等。其中斧劈纹、莲花纹、千层纹较佳，豆瓣纹次之。

龟纹石以其纹凹凸表现景观，能将山岳景观缩影在咫尺之内，大有"我持此石归，袖中有东海"的感觉。偏爱收藏景观石的藏家对龟纹石情有独钟，爱之如子。一块精品龟纹石表现出的山型景观能将绵绵山峦、林立峭壁、沟壑、平台等展现的惟妙惟肖，能将三山五岳的风光尽收其中，如临其境。其景观妙处不愧为大自然鬼斧神工的精品之作。龟纹石养护上无需上蜡，保持自然原貌最好。可用草酸或稀盐酸清洗一遍石体土锈，以后常保持清洁让其自然风化，更能表现自然风光。不要用电刷或手抚摸至其所谓的有层包浆，龟纹石属山石，不能追求包浆，有包浆反而失去大自然的原生状态，外地水石有的可有层包浆。灰白龟纹石和龟纹石与白色的方解石及结核石共生龟纹石种分布较广，皖山山脉大多山上都有。红龟纹石属于稀有石种，仅在南塘老鹰山几米见方的地方才有，周围几十千米都难以寻找到此类石种。

八公山模树石在书中又叫裟婆石、醒酒石。在远古时代的地质活动中，铁、锰的氧化物在地下水及温度和压力的共同作用下，沿岩石的节理、裂隙及层理等空隙处渗透，历经长期沉淀结晶形成板岩上的画面。（多呈现松树形、柏枝形或树与草密集成群的图案）。由于其形状很像树枝状植物化石，故有"假化石"之称。受沉淀物多寡及含铁锰元素的影响，其图案呈黑、红、黄、青、灰等多种色彩，犹如天然彩墨石画。以紫金石为底，形成的模树石所表现的画面，好似美术大师结合中西美术画法创作的精美图画，巧夺天工，绝妙无比。其中不但有立体感，墨色的浓淡干湿，都表现得淋漓尽致。一幅模树石形成的精品画面，更能引人入胜，浮想联翩，但也极其罕见难求。模树石在养护上防水、防尘，用塑料薄膜覆盖画面，装在订做的盒子里即可。模树石精品目前在淮南仅有两方，"山雨欲来"、"楼兰古色"（图15、图16）。

八公山瓜子石是石体上布满像瓜子片状的颗粒，因其外形像西瓜子，因而玩石的藏家把它命名为瓜子石。今后中国观赏石协会将对它也许会有科学的命名。瓜子石上的瓜子片有黑、黄、白、红四种颜色，石体表面有黄、灰两种底色。成分以碳酸钙为主，硬度在摩氏3°以上。孤立成形的瓜子石可做为奇石赏玩，将瓜子石表皮打磨经过抛光处理，有种远古化石的感觉。瓜子石奇石极少，当地奇石藏家手里仅有几块象形石，很少有带洞的或景观石。有洞和山峰的少之又少，鉴赏瓜子石要看瓜子片是不是凸出感强，瓜子片要大而独立，瓜子与底色要对比清晰，石型要完整。经过打磨的瓜子石有时也能表现出画面，但不能列入天然奇石行列。当地用瓜子石雕刻砚台、茶海、镇纸等工艺品远销海内外。其瓜子颗粒状是怎样形成的，是否是石友说的百合化石还有待考证。在养护上，可上无色石蜡养护。瓜子

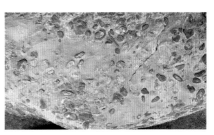

图17 "砚台" 八公山瓜子石 徐光华藏品

图18 八公山鱼籽石 徐光华藏品

石分布在南塘及丁家山后面老虎洞一带山上（图17）。

八公山鱼籽石是石体上布满像鱼籽状的圆形颗粒而得名的石种。也有鱼籽瓜子共生石，鱼籽晶莹透明，成分是方解石。该石以山型景观为主，若出象形的鱼籽石可列为精品。鱼籽石其底色有绛红和黄色，底色与鱼籽浑然一体形成画面石，画面耐人寻味极其漂亮深奥而被玩家收藏。因其石种稀少，具有观赏价值的鱼籽石世面极少。在养护上，可上无色石蜡养护。鱼籽石也分布在南塘及丁家山后面老虎洞一带山上。有瓜子石的山上也出鱼籽石。瓜子石和鱼籽石也是淮南特有石种（图18）。

八公山皖螺石，八公山皖螺石与灵

安徽淮南观赏石（连载二）

文：周保友 Author: Zhou Baoyou

Ornamental Stone of Huainan Region, Anhui Province (Serial II)

紫金石的观赏特征

紫金石属于彩石类观赏石。其分类有景观石、象形石、画面石等。景观石气势伟岸、意境深远、夺人心弦；象形石形神兼备、奇巧逼真；画面石画面布局匪夷所思，图案清晰，对比明快，有山水，人物，有大海之日出，有黄河之咆哮，尤如艺术大师独具匠心之作。紫金石色彩鲜艳，形象生动，石体上贯透紫金带、黄金带，石体表面有橙晕、包青、金线、蚰蜒纹、月白纹等天然俏色。紫金石不能过分要求石形瘦、皱、漏、透，但石体正面要突出紫金石色彩的特点。石体上紫带贯通，如行云流水，柔和飘逸，其中纯紫间有金色条纹最为高雅，紫中包青带有蚰蜒纹更是难得的精品。

若皖螺石与紫金石共生形成具有人文气息的紫金石，将会达到极品奇石的价值，可被列为国宝级奇石。紫金石纹理和石皮是鉴别天然原石的重要因素，纹理清晰有手感，石筋突出；但筋不能多，格不能乱；石筋自然有序，格纹点缀画面，都是上品紫金石。

紫金石开发出的紫金石工艺品精美绝伦，紫金石工艺品根据东方的神话传说有龙龟献宝、鱼跃龙门、紫金龙印、紫金砚等。紫金石工艺品不仅雕刻精细，其对紫金石的俏色利用要有创意并恰到好处方为精品。

龟纹石，八公山龟纹石属石灰岩，又叫碳酸岩，主要成分是碳酸钙。岩石剥离出的石体经过风化侵蚀和地下溶蚀，热胀冷缩，在石体表面形成纵横交错，似龟纹裂状而得名。外地也有该石种，如灵璧称其为莲花石。辽宁草花石与其颜色不同但纹理相近。

龟纹石品种有

一、红色龟纹石，地质成分是含泥质和赤铁矿较多的泥质石灰岩。方解石含赤铁矿形成的红龟纹石表面看去有玉化和玛瑙化的感觉（图12）。

二、灰白龟纹石，地质成分是可直接烧制石灰的含碳酸钙石灰岩（图13）。

三、龟纹石与白色的方解石及结核石共生龟纹石种（图14）。

以上石种红龟纹石罕见，大多是灰白色龟纹石。由于石体经过风化侵蚀和

图12 周保友藏品

图13 灰白龟纹石 周子俊藏品

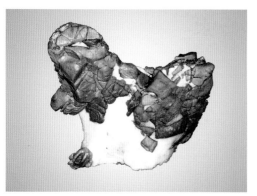

图14 "云雾山" 周子俊藏品

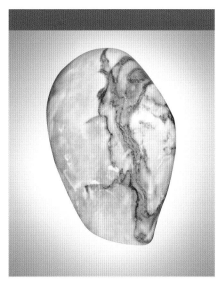

青田人物画面石

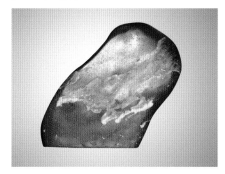

青田飞禽画面石

名贵石材，都产在山口。

走在山口乡的大街上，两旁村民家家锯石、户户雕刻，到处都是手工作坊。全村2000多户，几乎全都从事青田石雕业，产值占全乡工业产值90%以上。现在山口建有两个大型青田石交易市场，是国内外客商购石的重要场所。山口是青田石雕的发祥地，历代名家辈出，当今雕刻大师也大多出于此，也有大师辟有展室，珍贵藏品非常丰富。

青田石的开发与兴盛

中国古来治印，或以金属铸造，或以硬质材料琢磨。所谓文人治印，以软质佳石为纸，以刀代笔，尽显文人笔意情趣。在中国四大印石中，最早选用青田灯光冻石，始于元代赵孟頫，兴于明代文彭。青田石到清代，成为帝王的贡品、文人的至爱。当代的青田石得到极大的发展，成为印石爱好者争相寻找的藏品。

青田石的品种与收藏

根据新编《青田县志》，将青田石分为10大类、108个品种，颜色有红、黄、蓝、白、绿、黑等20多色，可谓品种繁多、色彩纷呈，美不胜收。青田石中最珍贵的品种有：

灯光冻

灯光冻石呈微黄色，细腻纯净，半透明如凝冻，光照灿若灯辉而得名。灯光冻扬名于明代，为青田石中极品。

封门清

封门清色淡青，产于封门矿洞，质细微透，纯净了无杂质而得品。封门清石材丰富，冻而清者珍贵。

蓝青田

蓝青田呈宝蓝色，质细腻而以色异而珍贵。蓝青田时有呈点状分布于浅色石料中，称为蓝星青田，别具风味。另有一种蓝钉青田，其钉硬度大，难奏刀，与蓝星青田不是同一个品种。

蜜蜡黄

蜜蜡黄色如蜜蜡，深浅各异，质地细腻，一般不透明或微透明，以质色取胜。

五彩冻

五彩冻一般由多种颜色组成，各种颜色过渡协调，同一颜色中又有层次变化。五彩冻色彩绚丽、质地细腻、行刀流畅，为青田珍品。

青田石收藏

硬度： 摩氏2~3°为佳，软易奏刀，方可如笔意游走。
质地： 温润细腻、冻若凝脂为上。
色泽： 单色纯净，多色协调，明亮光泽最佳。
纹理： 纹理奇特形成图案，有意韵者珍贵。

青田石产量大、品种多，精致的藏品尚有存量，升值空间很大，藏家要深入农户耐心寻访，必定会有收获。

【未完待续】

青田巧雕

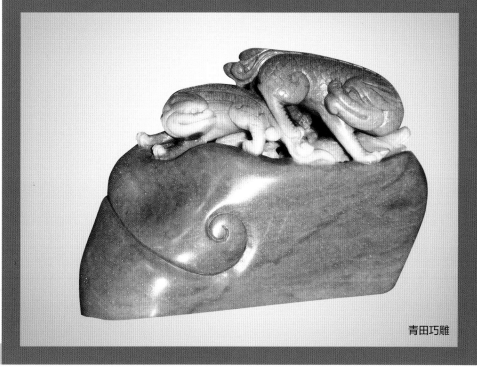

青田巧雕

China Scholar's Rocks 赏石中国

印章石集萃

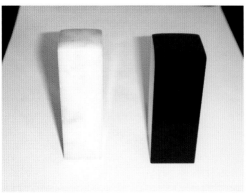

青田黑白石

农用汉制刻图书，印甚古。"山农为煮石山农的简称，是元代画家王冕的号，花药石即叶腊石，图书乃印章别称。赵孟頫、王冕在印章中的贡献，为明清印学起到推动作用。

明清以来，随着文人学者大力治印，论著增多。尤其是印章边款的创作和软质印石的使用，使我国印学成为独立的艺术门类，达到继两汉后的又一个艺术巅峰。

明代周应愿在《印说》中写道："文也，诗也，书也，画也，与印一也。"这种"印与文诗书画一体说"，将印提升到最高的审美境界。明代金光先《印章论》说："夫刀法贵明笔意，盖运刃如运笔。"明文学巨子王世贞讲："论印不于刀而于书，犹论字不以锋而以骨力，非无妙然。必胸中先有书法，用能迎刃而解。"明朱简《印经》说："印先字，字先章，章则具意，字则具笔。刀法者，所以传笔法也。刀法浑融，无迹可寻，神品也；有笔无刀，妙品也；有刀无笔，能品也；刀笔之外，而有别趣者，逸品也；有刀锋而似锯牙瘫股者，外道也；无刀锋而似铁线墨猪者，庸工也。"以上大家"笔意说"的精妙理论，使治印进入了更高的艺术殿堂。

明苏宣在《苏氏印略序》中说："知世不相沿，人自为政，如诗不法魏晋也，而非复魏晋；书不法钟王，而非复钟王。始于摹拟，终于变化。"清初学者周亮工在《书陆汉标印谱前》中说："陆汉标以予言为是，故任印能运己意。能运己意而复妙得古人意，此汉标之所以传也。"苏宣与周亮工提出的"入古出新"和"运己意说"，是对明代治印，出现一味刻意模仿，而失去自我的警示。为治印一道开辟了更为广阔的空间。

晚清印学大师赵之谦在其《苦兼室论印》中说："刻印以汉为大宗，胸有数百颗汉印，则动手自远凡俗。然后随功力所至，触类旁通，上追钟鼎法物，下及碑额造象，迄于山水花鸟，一时一事，觉无非印中旨趣，乃为妙悟。印以内为规矩，印以外为巧，规矩之用熟，则巧生焉。"赵之谦"印外求印说"，是讲治印功力在印外，功力所至，触类旁通。这种高瞻远瞩的见解，极大地丰富了印学宝库。

印学的流派与四大名印石的形成

由于印学的成熟、软质印料的普遍使用、书画诗文与印章结合得越发紧密、文人学者操刀治印成为时尚。经过明清的发展，出现了许多高手，形成艺术流派，主要有吴门派（代表人物：文彭）、皖派（代表人物：何震）、浙派（西泠四家：丁敬、奚冈等），后四家陈鸿寿（号曼生）等，以及赵之谦、吴俊卿（字昌硕）等。这些著名大师，共同缔造了印学的辉煌。

中国文房原为三大名印石，有青田石、寿山石、昌化石，至清代始成形。20世纪70年代巴林石被重新发现，列入文房名石中，中国文人四大名印石遂成定论。

1. 寻访青田石

乘飞机降落在浙江省温州市，驱车沿高速公路北上行驶不足1小时，就可以到达青田县城。

青田石名称

青田县始建于唐景云二年（711）。青田县城位于鹤城镇，是一个北依青田山，南临瓯江的狭长区域，最窄处只有几十米宽，却很绵长，因背后是青田山而得名。青田山又名太鹤山，所以县城又称为鹤城。太鹤山是道教圣地第三十洞天，古松、奇石、层峦、溪谷，风光旖旎，名胜颇多，是观光好去处。青田境内盛产易于奏刀的印石，石从县名，遂有青田石之称。

青田石之乡山口

青田石的主产地在山口乡，与青田城县城隔瓯江而望。近年来江上建起三座彩虹桥，往来两地十分方便。山口位于瓯江南岸的灵溪之畔，距县城18km，山苍水碧、景色迷人，也是探宝的好地方。山口山中出产印石的矿带，长约5km，厚度只有30m左右。由于矿脉分布的位置不同，形成5大采区，共有50多条采矿巷道。著名的灯光冻、封门清、五彩青田等

图19 "避雨" 八公山皖螺石 黄士勇藏品

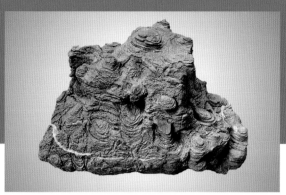

图20 红皖螺石 黄士勇藏品

图21 "国宝" 周保友藏品

璧皖螺石外观差不多，八公山皖螺石其原石为凹凸形鳞状线体构成，外表如鳞片盘错，石面上似排列着无数条头尾隐现的龙身形体。八公山皖螺石多为线状图纹，金钱状深纹的很少，线状皖螺轻轻敲击产生清脆悦耳的声音是八公山皖螺石与灵璧皖螺石的主要区别。八公山皖螺石石质坚硬，色彩悦目，品种有红皖螺、青皖螺及皖螺石与紫金石共生等。皖螺石形状有山形避雨，天池、过桥、云头雨脚及象形石等，但带洞的较少。青皖螺石多出在李家冲，闪家冲一带山上，红皖螺石出在寿县张管一带山脚下面。八公山皖螺石也是外地石友追捧的好石种。在养护上可上蜡也可不上蜡（图19、图20）。

八公山砂砾岩石，该石不仅能出造型石而且石体上有的还能形成画面可欣赏。据观赏石鉴评师培训教材中介绍，岩浆岩、沉积岩和变质岩的任何非均匀性的构造和结构都能在岩石中构成图案和纹理。八公山砂砾岩石是沉积岩先后沉积的颗粒大小、成分和颜色的差异不均匀现象的砾状结构。八公山砂砾岩石与灵璧品种石类似，不太了解的都认为是灵璧品种石，砂砾岩石没有被石友普遍开发和收藏，外地和本地石友只做为一个石种收藏。淮南奇石藏家手中精品很少。砂砾岩石在养护上一般需上无色石蜡养护。八公山砂砾岩石分布在闪家冲、李家冲、龙茅冲一带山上（图21）。

八公山虎皮石，是岩石经历断裂褶曲与含氧化铁、氧化硅的方解石浸染沉积共生形成犹如虎皮的斑纹而得名的石种。虎皮石质地坚硬，黑灰色带有橘红条纹，精品不多。虎皮石分布在南塘半截山（图22、图23）。

八公山当地石种还有根据石皮及色彩被当地奇石玩家命名的梨皮石、彩玉石、墨石、姜石、钟乳石等。在八公山南塘发现的腕须动物遗迹化石和古生物骨骼化石已被科研部门列为重点保护对象（图24）。

在淮南观赏石石种中，唯有八公山的紫金石是全国独有的石种，其品种多样且储藏量较大。定远、山东也有和紫金石外观相似的石头，但其色彩不如八公山的紫金石丰富。紫金石在南宋时期曾被开发，达官贵人、文人墨客，当成供石或雕刻成山籽摆件，置于几案厅堂，或装点园林家居。据《淮南子·览冥训》记载："往古之时，四极废，九州裂，天不兼覆，地不周载，火爁炎而不灭，水浩洋而不息，猛兽食颛民，鸷鸟攫老弱。于是女娲炼五彩石以补苍天，断鳌足以立四极，杀黑龙以济冀州，积芦灰以止淫水。"相传女娲补天炼的五彩紫金石遗落人间，化为紫金山，坐落在八公山脉。在淮南的传说中另有一段佳话，相传宋太祖赵匡胤与西周大将余洪在八公山下厮杀，余洪被困在八公山西麓楗笼冲。太祖赵匡胤久攻不下，后下令用火焚山。余洪大军全部覆没，山上石头也被烧成红色，当地农民把这红色紫金石纷纷说成是宋太祖火烧余洪烧成的，作为传说，留传至今。神话故事《封神演义》书中石矶娘娘的道场也在八公山中，现已改建为道教活动场所，常年香火旺盛。紫金石的文化传说伴随着淮南地方文化充填着淮南的精神文明建设，丰富着淮南人民的精神生活。

每当节庆期间，淮南市政府和八公山区政府与淮南奇石协会都会举办"淮南市紫金石工艺暨奇石展"，开发淮南的旅游资源及奇石文化。在全国各地举办的大型观赏石展中都有安徽淮南紫金石参展并获奖项。现全国各地大都知道安徽淮南紫金石。因而淮南八公山紫金石的开发具有很大的价值空间。今后安徽淮南紫金石将会被中国观赏石协会编入《当代中国观赏石谱》中，并向中华名石努力，绽放其独特的风采。

图22 杨金社藏品

图23 "恐龙" 周保友藏品

图24 周保友藏品

中国唐苑

THE RESURRECTION OF CULTURE BELONGING TO TANG DYNASTY STARTS FROM HERE

唐苑
中国唐苑